CAMBRIDGE STUDIES IN
ADVANCED MATHEMATICS: 54

LECTURES ON VECTOR BUNDLES

Already published

1 W.M.L. Holcombe *Algebraic automata theory*
2 K. Petersen *Ergodic theory*
3 P.T. Johnstone *Stone spaces*
4 W.H. Schikhof *Ultrametric calculus*
5 J.-P. Kahane *Some random series of functions, 2nd edition*
6 H. Cohn *Introduction to the construction of class fields*
7 J. Lambek & P.J. Scott *Introduction to higher-order categorical logic*
8 H. Matsumura *Commutative ring theory*
9 C.B. Thomas *Characteristic classes and the cohomology of finite groups*
10 M. Aschbacher *Finite group theory*
11 J.L. Alperin *Local representation theory*
12 P. Koosis *The logarithmic integral I*
13 A. Pietsch *Eigenvalues and s-numbers*
14 S.J. Patterson *An introduction to the theory of the Riemann zeta-function*
15 H.J. Baues *Algebraic homotopy*
16 V.S. Varadarajan *Introduction to harmonic analysis on semisimple Lie groups*
17 W. Dicks & M. Dunwoody *Groups acting on graphs*
18 L.J. Corwin & F.P. Greenleaf *Representations of nilpotent Lie groups and their applications*
19 R. Fritsch & R. Piccinini *Cellular structures in topology*
20 H Klingen *Introductory lectures on Siegel modular forms*
22 M.J. Collins *Representations and characters of finite groups*
24 H. Kunita *Stochastic flows and stochastic differential equations*
25 P. Wojtaszczyk *Banach spaces for analysts*
26 J.E. Gilbert & M.A.M. Murray *Clifford algebras and Dirac operators in harmonic analysis*
27 A. Frohlich & M.J. Taylor *Algebraic number theory*
28 K. Goebel & W.A. Kirk *Topics in metric fixed point theory*
29 J.F. Humphreys *Reflection groups and Coxeter groups*
30 D.J. Benson *Representations and cohomology I*
31 D.J. Benson *Representations and cohomology II*
32 C. Allday & V. Puppe *Cohomological methods in transformation groups*
33 C. Soulé et al *Lectures on Arakelov geometry*
34 A. Ambrosetti & G. Prodi *A primer of nonlinear analysis*
35 J. Palis & F. Takens *Hyperbolicity, stability and chaos at homoclinic bifurcations*
37 Y. Meyer *Wavelets and operators I*
38 C. Weibel *An introduction to homological algebra*
39 W. Bruns & J. Herzog *Cohen-Macaulay rings*
40 V. Snaith *Explicit Brauer induction*
41 G. Laumon *Cohomology of Drinfield modular varieties I*
42 E.B. Davies *Spectral theory and differential operators*
43 J. Diestel, H. Jarchow & A. Tonge *Absolutely summing operators*
44 P. Mattila *Geometry of sets and measures in Euclidean spaces*
45 R. Pinsky *Positive harmonic functions and diffusion*
46 G. Tenenbaum *Introduction to analytic and probabilistic number theory*
47 C. Peskine *An algebraic introduction to complex projective geometry I*
50 I. Porteous *Clifford algebras and the classical groups*
51 M. Audin *Spinning Tops*
54 J. Le Potier *Lectures on Vector bundles*

Lectures on Vector Bundles

J. Le Potier

Université Paris VII

Translated by A. Maciocia

CAMBRIDGE
UNIVERSITY PRESS

PUBLISHED BY THE PRESS SYNDICATE OF THE UNIVERSITY OF CAMBRIDGE
The Pitt Building, Trumpington Street, Cambridge CB2 1RP

CAMBRIDGE UNIVERSITY PRESS
The Edinburgh Building, Cambridge CB2 2RU, United Kingdom
40 West 20th Street, New York, NY 10011-4211, USA
10 Stamford Road, Oakleigh, Melbourne 3166, Australia

First published 1997

Typeset in Computer Modern

A catalogue record for this book is available from the British Library

Library of Congress cataloguing in publication data

Le Potier, Joseph, 1944–
Lectures on vector bundles / J. Le Potier ; translated by A. Maciocia.
p. cm. – (Cambridge studies in advanced mathematics ; 54)
Includes bibliographical references (p. –) and index.
ISBN 0 521 48182 1 (hardback)
1. Vector bundles. I. Title. II. Series.
QA612.63.L4 1997
516.3′5–dc20 96-13367 CIP

ISBN 0 521 48182 1 hardback

Transferred to digital printing 2004

TAG

Contents

Preface

This work consists of two courses treating the moduli spaces of vector bundles. The first part is an elementary DEA course which we have given at the Université Denis Diderot (Paris VII) in the Spring of 1991: the subject tackled is the classification of vector bundles on algebraic curves, the construction of the moduli spaces of stable bundles and the first properties of these moduli spaces. In particular, one will find the smoothness criteria and the existence conditions of stable bundles of given rank and degree. It consists of some reminders of facts from algebraic geometry and algebraic topology, the construction, in the case of curves, of the Hilbert-Grothendieck scheme of coherent quotient sheaves of a given bundle, and to treat this we also give a succinct description of Mumford's geometric invariant theory.

The second part is a more advanced course given at the school "Vector Bundles on Surfaces" organized by EUROPROJ (European projective geometry network) and CIMI (Centre International pour les Mathématiques et l'Informatique) at Nice Sophia-Antipolis (14–18 June 1993). The subject we treated here centred on the structure of the moduli space of semi-stable sheaves on the projective plane. In particular, we establish the existence conditions of semi-stable sheaves of given rank and Chern classes and deal with the question of irreducibility and describe the Picard group of the moduli spaces. The construction of the moduli space of semi-stable sheaves is sketched in the more general situation of projective algebraic surfaces: it is fairly similar to the construction we shall see in the first part for the case of curves once we have established the fact that the family of semi-stable sheaves with fixed Hilbert polynomial is bounded; in fact, it generalizes without much dificulty to higher dimensional varieties. The fact that this family is bounded rests on the theorems dealing with restriction of semi-stable sheaves to curves which lead to Bogomolov's Theorem: the statement we shall show here is due to Flenner [20]. This course assumes knowledge of Chern classes of vector bundles and of coherent algebraic sheaves and all the fundamental results of algebraic geometry, notably Serre's Theorems A and B, the Finiteness Theorem and the Riemann-Roch formula, some of which is dealt with in the first part and for the rest the reader could consult, for example, Hartshorne's book [31].

I would like to thank Christoph Sorger who delivered and revised the first version of chapter 6 which deals with Mumford's geometric invariant theory. The enormous work of translation was carried out by Antony Maciocia. I would like to express my satisfaction and my gratitude to him for the result.

Paris, 22nd October 1995.
J. Le Potier

By algebraic variety we mean a separated scheme of finite type over $\mathbb{C}$ and by points we shall always mean closed points.

Part I

Vector Bundles on Algebraic Curves

Introduction

The following is the text of a course given at the Université Denis Diderot (Paris VII) in Spring 1991. Although it was aimed at students who had already followed a third level Algebraic Geometry course, in chapters 1 and 2, we review the fundamental statements and particularly those regarding the cohomology of coherent algebraic sheaves, the Serre Duality Theorem and the Riemann-Roch formula. This chapter contains only a few proofs and we refer the reader for more details to basic algebraic geometry texts such as the books of Hartshorne [31] and Shafarevich [57]. Properties connected with direct images and properties of flatness (the Semi-continuity Theorem and the criterion for flatness) are dealt with in chapter 4. In this chapter, we shall also describe the construction of the Hilbert-Grothendieck scheme of coherent sheaves with fixed Hilbert polynomial which are quotients of a given vector bundle over a projective algebraic curve.

The subject of this course is the problem of the classification of algebraic vector bundles on a connected, smooth, projective algebraic curve X of genus g. To simplify matters, we restrict ourselves to the case where the base field is $\mathbb{C}$, which allows us to view the degree as a topological invariant and to obtain the topological classification of such bundles immediately: in effect, they are classified by their rank and degree (cf. chapter 3). This allows us to study algebraic vector bundles with given rank r and degree d. This family is not bounded (cf. chapter 5) and so, to parametrize vector bundles by algebraic varieties, it is natural to introduce the notions of stability and semi-stability. Other bundles are classified by reducing them using a Harder-Narasimhan filtration.

One finds very quickly that, for large enough d, the set of isomorphism classes of semi-stable bundles of rank r and degree d can be identified with the set of orbits under a natural action of the group $\mathrm{SL}(H)$ on an invariant open subset Ω of the Hilbert-Grothendieck scheme $\mathbf{Hilb}^{r,d}(H\otimes \mathcal{O}_X)$ of coherent quotient modules of the trivial bundle $H \otimes \mathcal{O}_X$, where H is a vector space of dimension $\chi = d + r(1-g)$. The problem then is to equip the quotient with the structure of an algebraic variety. It is here that Mumford's geometric invariant theory arises (cf. [55], [47], [50]). This Hilbert scheme embeds as a closed sub-variety of projective space and the open subset Ω is actually the open set of semi-stable points under the $\mathrm{SL}(H)$ action. Mumford's theory shows that this open subset has a good quotient $\mathbf{M}(r,d)$ and that this good quotient is a projective algebraic variety, the points of which correspond to certain equivalence classes of semi-stable bundles: this is the notion of S-equivalence, which is defined in terms of Jordan-Hölder filtrations. For stable bundles, these equivalence classes reduce to isomorphism classes, which gives the set of isomorphism classes of stable vector bundles of fixed rank and degree the structure of a quasi-projective variety $\mathbf{M}^s(r,d)$.

The above technique, now very classical, goes back to the work of Mumford and Seshadri (cf. [55], [47], [51]). However, the embedding that we have chosen in order to regard the Hilbert scheme $\mathbf{Hilb}^{r,d}(H\otimes\mathcal{O}_X)$ as a closed sub-variety of a projective space differs from the one introduced by these authors: we make use of the polarization which Grothendieck introduced in its initial construction [29], which results in proofs which are very much easier and without doubt more natural (cf. chapter 7). It has the other great advantage that it can be generalized to singular curves without much change: and, moreover, the method lends itself to the construction of moduli spaces of semi-stable sheaves on varieties of larger dimension without major modification, as one can see by studying, for example, Simpson's article [58].

Chapter 8 is devoted to proving the theorems of existence, irreducibility and smoothness of the open subset of stable bundles. The smoothness criterion for the points of the Hilbert scheme $\mathbf{Hilb}^{r,d}(H, \mathcal{O}_X)$ is obtained following an idea of Grothendieck [29] and involves studying the possible extensions of the quotients parametrized by fat points, a method that goes by the name of the criterion of formal smoothness. The use of the criterion of formal smoothness here seems to be unavoidable because, in a neighbourhood of a point, the Hilbert scheme arises as an intersection of a finite family of determinantal varieties, so it is not easy to determine their codimension. We have no references for the theorems of chapter 8; the method used is the adaption of an exact sequence which J.-M. Drézet and the author introduced to study the existence of stable bundles on the projective plane [15] and which is explained more fully in Part II. We have included a problem at the end of Part I which is concerned with the study of algebraic vector bundles on elliptic curves and is obviously inspired by the article of M.F. Atiyah [1].

1

Generalities

In this chapter, we shall review some of the basic facts about vector bundles. The main reference is the book of Shafarevich [57], Chapter VI. We shall assume that the reader has already met the notion of an algebraic sheaf or $\mathcal{O}_X$-module. We recall the main operations on vector bundles.

1.1 Definition

Let X be an algebraic variety[1]. A (*complex*) *linear fibration* over X is given by an algebraic variety E, a surjective morphism $p : E \to X$ of algebraic varieties and, for each point $x \in X$, a complex vector space structure on the fibre $p^{-1}(x)$. The variety E is called the *total space* of the fibration and the fibre over x is denoted by E_x. Given two fibrations $p : E \to X$ and $p' : E' \to X$, a morphism of varieties $f : E \to E'$ is a map of linear fibrations if it is compatible with the projections p and p', that is, $p' \circ f = p$, and if, for each $x \in X$, the induced map $f_x : E_x \to E'_x$ is linear. The bundle $X \times \mathbb{C}^r \to X$ given by projection to the first factor is called the *trivial fibration* of rank r. For each open set $U \subset X$, we write $E|_U$ for the fibration $p^{-1}(U) \to U$ given by restriction to U.

An *algebraic vector bundle* of rank r on X is a linear fibration $E \to X$ which is locally trivial in the following sense: for each point $x \in X$ there exist an open neighbourhood U of x and an isomorphism of fibrations

$$\varphi : E|_U \longrightarrow U \times \mathbb{C}^r.$$

[1] The base field is $\mathbb{C}$. In this chapter we will assume that the varieties are reduced but most of the results apply to the non-reduced case.

1.2 Transition functions

Such an isomorphism is called a *local chart* or a *trivialization* of the vector bundle which, abusing the notation, we denote by E. If $\varphi_i : E|_{U_i} \to U_i \times \mathbb{C}^r$ and $\varphi_j : E|_{U_j} \to U_j \times \mathbb{C}^r$ are local charts over the open subsets U_i and U_j respectively then the change of chart defined over $U_{i,j} = U_i \cap U_j$ by

$$\varphi_i \circ \varphi_j^{-1} : U_{i,j} \times \mathbb{C}^r \longrightarrow U_{i,j} \times \mathbb{C}^r$$

takes the form $(x, v) \mapsto (x, g_{i,j}(x)v)$, where $g_{i,j} : U_{i,j} \to \mathrm{GL}(r, \mathbb{C})$ is a map of algebraic varieties. The maps $g_{i,j}$ are called *transition functions.* They satisfy the following conditions:

(a) $g_{i,j} = \mathrm{id}_{\mathbb{C}^r}$ over the open subset U_i,

(b) over the intersection $U_{i,j,k} = U_i \cap U_j \cap U_k$ we have $g_{i,k} = g_{i,j} g_{j,k}$.

1.3 The vector bundle associated to transition functions

Let $\mathfrak{U} = (U_i)_{i \in I}$ be an open cover of X. For each $(i, j) \in I^2$, suppose that we are given a map $g_{i,j} : U_{i,j} \to \mathrm{GL}(r, \mathbb{C})$ satisfying conditions (a) and (b) above. We consider the quotient

$$E = \Big(\coprod_{i \in I} U_i \times \mathbb{C}^r\Big)/R$$

under the equivalence relation which identifies the points $(x, v) \in U_i \times \mathbb{C}^r$ with $(x', v') \in U_j \times \mathbb{C}^r$ when $x = x'$ and $v' = g_{i,j}(x)v$. This set is given the quotient topology. We then have a continuous projection $p : E \to X$ and a homeomorphism

$$E|_{U_i} \xrightarrow{\sim} U_i \times \mathbb{C}^r$$

over U_i. We can then give the topological quotient space E the structure of an algebraic variety (separated and of finite type) induced from the algebraic variety structure of $U_i \times \mathbb{C}^r$. Over $U_{i,j}$, the structures induced from $U_i \times \mathbb{C}^r$ and from $U_j \times \mathbb{C}^r$ coincide. One can also carry over the vector space structure of the fibres: this vector space structure on E_x does not depend on i. Thus, we obtain a vector bundle $E \to X$ and by definition the above isomorphism gives a trivialization. So from a system of transition functions we have constructed a vector bundle over X.

1.4 Maps of vector bundles

Suppose that we are given two vector bundles E and F of ranks r and s, respectively, over X. Then a map from E to F is a map of the underlying linear fibrations. If $\varphi : E|_U \to U \times \mathbb{C}^r$ and $\psi : F|_U \to U \times \mathbb{C}^s$ are local charts for the bundles E and F over the same open subset U then the maps $\tilde{f} = \psi f \varphi^{-1} : U \times \mathbb{C}^r \to U \times \mathbb{C}^r$ are the local expressions of f in the charts φ and ψ, and are written as

$$(x, v) \mapsto (x, g(x)v),$$

where $g : U \to L(\mathbb{C}^r, \mathbb{C}^s)$ is a map of algebraic varieties from U to the vector space $L(\mathbb{C}^r, \mathbb{C}^s)$ of linear maps $\mathbb{C}^r \to \mathbb{C}^s$.

Proposition 1.4.1 *Let $f : E \to F$ be a map of algebraic vector bundles and let x_0 be a point such that f_{x_0} is invertible. Then there exists an open neighbourhood U of x_0 such that $f|_U : E|_U \to F|_U$ is an isomorphism.*

Proof It suffices to prove this for the local expression for f. The hypotheses imply that $g(x_0)$ is invertible. Since g is continuous, $g(x)$ is invertible on an open neighbourhood U of x_0 and we can use this to construct the inverse of f. □

1.5 Associated bundles

Given two bundles E and F over X, the *direct sum* of E and F is defined, as a set, by

$$E \oplus F = \coprod_{x \in X} (E_x \oplus F_x).$$

To equip this set with the structure of an algebraic variety we start by choosing local charts $\varphi_i : E|_{U_i} \to U_i \times \mathbb{C}^r$ and $\psi_i : F|_{U_i} \to U_i \times \mathbb{C}^s$ over the same open set U_i. Consider the bijections

$$(E \oplus F)|_{U_i} \to U_i \times (\mathbb{C}^r \oplus \mathbb{C}^s)$$

associated to φ_i and ψ_i and carry over the algebraic structure from the right hand side using this bijection. To see that this structure is independent of the choice of charts, it suffices to calculate the change of charts for $E \oplus F$: over the open set $U_{i,j}$ it is given by the transition functions $U_{i,j} \to \mathrm{GL}(\mathbb{C}^r \oplus \mathbb{C}^s)$ defined by the matrix

$$\begin{pmatrix} g_{i,j} & 0 \\ 0 & h_{i,j} \end{pmatrix},$$

where $g_{i,j}$ and $h_{i,j}$ are the transition functions of E and F respectively.

In the same way one can construct the bundles $\mathcal{H}om(E, F)$, $E \otimes F$, the dual bundle E^*, the exterior power $\Lambda^k(E)$ and the symmetric power $S^k(E)$.

1.6 Sub-bundles

Let $p : E \to X$ be an algebraic vector bundle of rank r over X. A *sub-bundle of rank m* of E is a sub-variety $F \subset E$ such that, for each $x \in X$, the intersection $F \cap E_x$ is a vector sub-space of E_x of dimension m and such that the induced fibration

$$p|_F : F \longrightarrow X$$

is locally trivial.

Example.
Let $X = \mathbf{P}_n(\mathbb{C})$ be n-dimensional projective space. We let $(x_0, \ldots, x_n)$ denote the homogeneous coordinates of a point $x \in X$. Consider the set of pairs $(x, v) \in X \times \mathbb{C}^{n+1}$ such that $v \in x$. This is a closed sub-variety H of $X \times \mathbb{C}^{n+1}$. Over the open set U of X defined by $x_i \neq 0$, W is defined by the equations

$$v_j = \frac{x_j}{x_i} v_i, \quad j = 0, \ldots, n.$$

The fibre H_x of $H \to X$ over a point x can be identified with the line x in $\mathbb{C}^{n+1}$. Then we have an isomorphism of fibrations

$$U_i \times \mathbb{C} \longrightarrow H|_{U_i}$$

given by $(x, t) \mapsto (x, v)$, where v is defined by $v_j = \frac{x_j}{x_i} t$ for $j = 0, \ldots, n$. It follows that H is a rank 1 vector bundle over X. This bundle is called the *Hopf bundle* and its dual is denoted $\mathcal{O}(1)$.

1.7 Quotient bundles

Let $F \subset E$ be a sub-bundle of an algebraic vector bundle E. Consider the family of vector spaces over X

$$E/F = \coprod_{x \in X} (E_x / F_x).$$

Any map of vector bundles $f : E \to G$ which vanishes on F factorizes uniquely through a map $\bar{f} : E/F \to G$.

Proposition 1.7.1 *There exists a unique vector bundle structure on $E/F \to X$ which satisfies the following universal property: for each map $f : E \to G$ which vanishes on F, the map $\bar{f}$ is a map of algebraic vector bundles.*

Proof Let x be a point of X. Over an open neighbourhood U of x we can find a vector sub-bundle S of $E|_U$ such that $F_x \oplus S_x = E_x$. Then the canonical map of vector bundles

$$F|_U \oplus S \longrightarrow E|U$$

is an isomorphism over the point x and so gives rise to an isomorphism over some open neighbourhood V of x by Proposition 1.4.1. This gives rise to a bijection $k : S|_V \to (E/F)|_V$ which is compatible with the vector space structures on the fibres. This identification gives rise to a vector bundle structure on $(E/F)|_V$. It is clear that the structure we obtain in this way over V is independent of the choice of S and of V, the induced structures match up. So we obtain a vector bundle structure on E/F which satisfies the given universal property because $\bar{f} \circ k$ is the restriction of f to $S|_V$ and, therefore, a map of vector bundles. □

Proposition 1.7.2 *Let $f : E \to F$ be a map of vector bundles. Suppose that the rank of f_x remains constant as x varies over X. Then* $\ker f$ *and* $\operatorname{Im} f$ *are sub-bundles of E and F respectively.*

Proof The question is local on X. Let x be a point of X. Choose sub-bundles K and S of $E|_V$ and L and T of $F|_V$ over a suitable open neighbourhood V of x so that $K_x = \ker f_x$, $L_x = \operatorname{Im} f_x$ and such that

$$K \oplus S = E|_V \quad \text{and} \quad L \oplus T = F|_V.$$

With respect to these direct sums, we write the map f as

$$f = \begin{pmatrix} a & b \\ c & d \end{pmatrix}.$$

Over the point x, the maps a, c and d are zero and b is invertible. By Proposition 1.4.1, b is an isomorphism over a neighbourhood of x. Then the kernel of f is given by the pairs $(u, v) \in K \oplus S$ such that $(c - db^{-1}a)u = 0$ and $v = -b^{-1}au$. It now follows that the map $c - db^{-1}a$ has constant rank which must be zero because it vanishes at x. Then the isomorphism $(u, v) \mapsto (u, v + b^{-1}au)$ maps the kernel of f to the first factor of the direct sum and so it is a sub-bundle. Similarly, the image of f is identified with the set of pairs $(\xi, \eta) \in L \oplus T$ such that $\eta = db^{-1}\xi$ and the isomorphism $(\xi, \eta) \mapsto (\xi, \eta - db^{-1}\xi)$ maps the image of f onto the first factor of the direct sum. Consequently, this image is also a sub-bundle[1]. □

Corollary 1.7.3 *Let $f : E \to F$ be a map of vector bundles. If f is surjective then* $\ker f$ *is a sub-bundle of E. If f is injective then* $\operatorname{Im} f$ *is a vector sub-bundle of F.*

Example.
Let X be a smooth algebraic sub-variety of $\mathbb{C}^n$ of dimension m defined by an equation $f = 0$, where $f : \mathbb{C}^m \to \mathbb{C}^k$ is a polynomial map. Then the differential $(x, v) \mapsto (x, d_x f(v))$ has constant rank over X. Its kernel is a sub-bundle of rank n of $X \times \mathbb{C}^m$ called the *tangent bundle* to X and denoted $T(X)$. If X' is another smooth sub-variety of $\mathbb{C}^{m'}$ then each map $\varphi : X \to X'$ induces a map of vector bundles $T\varphi : T(X) \to T(X')$.

Example.
More generally, let X be a smooth algebraic variety of dimension n. We can always find a cover of X by open affine sets U_i and for each i an isomorphism $\varphi_i : U_i \to U_i'$ to a closed sub-variety of $\mathbb{C}^{n_i}$. The *tangent space* at x to X is given by $T_x X = (\mathfrak{m}_x/\mathfrak{m}_x^2)^*$, where $\mathfrak{m}_x$ is the maximal ideal of the local ring of X at the point x. Consider the space

$$T(X) = \coprod_{x \in X} T_x X \longrightarrow X.$$

The isomorphism φ_i induces an isomorphism $T_x\varphi_i : T_x \to T_{\varphi_i(x)}(U_i')$. So we have a bijection

$$T(X)|_{U_i} \longrightarrow T(U_i')$$

defined by $(x, v) \mapsto (\varphi_i(x), T_x\varphi_i(v))$ which is linear in each fibre and allows us to transfer the vector bundle structure to $T(X)|_{U_i}$. Over the open set $U_{i,j} = U_i \cap U_j$, such structures defined by φ_i and φ_j coincide.

1.8 Sections

Let $p : E \to X$ be a vector bundle of rank r over an algebraic variety X. We define a *regular section* of E over an open set U to be a map $s : U \to E$ of algebraic varieties such that $p(s(x)) = x$ for all $x \in U$.

[1] The statement of Proposition 1.7.2 does not remain true for a non-reduced variety but the corollary is still true.

The set $\Gamma(U, E)$ of regular sections of E over an open set U can be given the structure of a module over the algebra $\mathcal{O}(U)$ of regular functions on U. The operations are defined by

$$(s+t)(x) = s(x) + t(x) \quad \text{and} \quad (\alpha s)(x) = \alpha(x)s(x)$$

for $s, t \in \Gamma(U, E)$ and $\alpha \in \mathcal{O}(U)$. In this way we obtain a sheaf of $\mathcal{O}_X$-modules $\mathcal{E} = \mathcal{O}_X(E)$ over X, locally isomorphic to $\mathcal{O}_X^r$: we say that the sheaf is *locally-free* of rank r.

Proposition 1.8.1 *The functor which associates the sheaf of modules of regular sections to a vector bundle E is an equivalence of categories between the category of algebraic vector bundles over X and the category of locally-free sheaves of finite rank on X.*

Proof The functor is fully faithful: in other words, for any two vector bundles E and F of ranks p and q respectively, the natural linear map

$$\mathrm{Hom}(E, F) \longrightarrow \mathrm{Hom}_{\mathcal{O}_X}(\mathcal{O}_X(E), \mathcal{O}_X(F))$$

is an isomorphism. The question is local on X and so one can suppose that $E = X \times \mathbb{C}^p$ and $F = X \times \mathbb{C}^q$. Any map $f : E \to F$ can be regarded as a map $X \to L(\mathbb{C}^p, \mathbb{C}^q)$ or as a morphism of $\mathcal{O}_X$-modules $\mathcal{O}_X^p \to \mathcal{O}_X^q$.

We now show that the functor is essentially surjective: given a locally-free sheaf $\mathcal{E}$ of rank r we need to construct a vector bundle E and an isomorphism $\mathcal{O}_X(E) \cong \mathcal{E}$. Let $\mathcal{E}_x$ be the module of germs of sections of $\mathcal{E}$ at a point x and consider the vector space $\mathcal{E}(x) = \mathcal{E}_x/\mathfrak{m}_x\mathcal{E}_x$, where $\mathfrak{m}_x$ is the maximal ideal of the local ring of X at the point x. This vector space has dimension r. Put

$$E = \coprod_{x \in X} \mathcal{E}(x) \longrightarrow X.$$

Let (U_i) be an open cover of X for which we have isomorphisms $\mathcal{O}_{U_i}^r \cong \mathcal{E}|_{U_i}$. These trivializations induce bijections $\varphi_i : U_i \times \mathbb{C}^r \cong E|_{U_i}$ and, over $U_{i,j} = U_i \cap U_j$, the map $\varphi_i \varphi_j^{-1}$ is a vector bundle isomorphism. So, if we use φ_i to transfer the local vector bundle structure of $X \times \mathbb{C}^r$ to E, then the structures we get over $U_{i,j}$ coincide. Hence we obtain a vector bundle. Sections of E over an open set U are given by sections s_i over $U \cap U_i$ which coincide over $U_{i,j}$. We deduce that there is an isomorphism of $\mathcal{O}_X$-modules $\mathcal{O}_X(E) \cong \mathcal{E}$. This completes the proof. □

1.9 The bundle associated to a representation

Let $f : Y \to X$ be a map of algebraic varieties and let $q : E \to X$ be a vector bundle of rank r on X. Consider the set $F = Y \times_X E$ of pairs (y, v) such that $f(y) = q(v)$. The natural projection $\mathrm{pr}_1 : F \to Y$ gives rise to a vector bundle of rank r on Y. The sheaf of sections of F is the inverse image of the sheaf of section of E as a sheaf of modules.

Consider the variety R of pairs (x, φ) consisting of a point of X and an isomorphism $\varphi : \mathbb{C}^r \to E_x$. Over an open set over which E is trivial, there is an isomorphism $R|_U \cong U \times \mathrm{GL}(r, \mathbb{C})$. This allows us to equip R with the structure of an algebraic variety on which the group $\mathrm{GL}(r, \mathbb{C})$ acts on the right. The action is fixed-point-free and the orbits can be identified with the fibres of the projection $R \to X$. Moreover, the variety X is a quotient of R in the categorical sense. This means that any map $R \to Z$ which is constant on the orbits factors through X. The fibration $R \to X$ is called the *frame bundle* of E. Consider the inverse image bundle $F = R \times_X E$. This bundle comes with a trivialization

$$R \times \mathbb{C}^r \cong F.$$

In fact, the bundle E is the quotient of $R \times \mathbb{C}^r$ by the action of the group $\mathrm{GL}(r, \mathbb{C})$ defined for $\gamma \in \mathrm{GL}(r, \mathbb{C})$, $\varphi \in R$ and $v \in \mathbb{C}^r$ by

$$(\gamma, (\varphi, v)) \mapsto (\varphi\gamma^{-1}, \gamma v).$$

Example.
The frame bundle of the Hopf bundle H can be identified with the projection map $\mathbb{C}^{n+1} \setminus \{0\} \to \mathbf{P}_n$. So H can be regarded as the quotient of $\mathbb{C}^{n+1} \setminus \{0\} \times \mathbb{C}$ by the equivalence relation which identifies (x, v) with $(\lambda^{-1}x, \lambda v)$ for $x \in \mathbb{C}^{n+1} \setminus \{0\}$, $v \in \mathbb{C}$ and λ a non-zero scalar.

Now consider a representation $\rho : \mathrm{GL}(r, \mathbb{C}) \to \mathrm{GL}(m, \mathbb{C})$. Then we can consider the quotient $E_\rho = R \times_\rho \mathbb{C}^m$ by the action of $\mathrm{GL}(r, \mathbb{C})$ defined by $(\gamma, (\varphi, t)) \mapsto (\varphi\gamma^{-1}, \rho(\gamma)(t))$. We then obtain a vector bundle of rank m over X which we call the *vector bundle associated to the representation* ρ. The module of global sections of E_ρ is isomorphic to the regular functions $s : R \mapsto \mathbb{C}^m$ satisfying the condition $s(\gamma x) = \rho(\gamma)s(x)$.

Example.
The bundle associated to the representation $\gamma \mapsto {}^t\gamma^{-1}$ is none other than the dual bundle E^*. The bundle associated to the representation $\gamma \mapsto \Lambda^k(\gamma)$ (respectively $\gamma \mapsto S^k(\gamma)$) is the exterior power $\Lambda^k(E)$ (respectively the symmetric power $S^k(E)$). If on projective space $\mathbf{P}_n = \mathbf{P}(V)$ we

consider the bundle $\mathcal{O}(k) = S^k(H^*)$ then this is the bundle associated to the representation $\gamma \mapsto \gamma^{-k}$. It follows that the space of sections of $\mathcal{O}(k)$ can be identified with the space of homogeneous polynomials $S^k(V^*)$ of degree k over V and that this space is zero if $k < 0$.

2

The Riemann-Roch Formula

In this chapter, we first review the definition of sheaf cohomology of abelian groups. In fact, we shall introduce the more general notion of hypercohomology of complexes of sheaves of abelian groups which are bounded below. Although not indispensable for what follows, this is a tool which is sometimes useful: it can be used, for example, to define algebraically the de Rham cohomology of algebraic varieties. Of course, this is the same method which allows us to define the derived and hyperderived functors of any left exact covariant additive functor $F : \mathcal{A} \to \mathcal{B}$ from an abelian category $\mathcal{A}$ to an abelian category $\mathcal{B}$, provided $\mathcal{A}$ has enough injectives. We refer the reader to chapter III of Hartshorne's book [31] for more details, or to Godement's book [23]. After this we recall some of the fundamental results about coherent algebraic sheaves: Serre's Theorems A and B on affine varieties and on projective varieties, the finiteness theorem on projective varieties and the Serre Duality Theorem [31]. This allows us to introduce the Euler-Poincaré characteristic and to establish the Riemann-Roch Theorem for curves which tells us how to calculate the degree of a coherent sheaf on a curve.

2.1 Cohomology

Let X be a topological space. Given a sheaf F of abelian groups[1] on X, we let F_x denote the group of germs of sections over a neighbourhood of x: this is just the inductive limit

$$F_x = \varinjlim_{U \ni x} F(U).$$

[1] All the sheaves in this section are sheaves of abelian groups

Recall that the category of sheaves of abelian groups is an abelian category. This implies that each morphism has a kernel, an image and a cokernel. Given two maps of sheaves a and b

$$F' \xrightarrow{a} F \xrightarrow{b} F''$$

we say that (a, b) is an *exact sequence* if $\ker b = \operatorname{Im} a$. This property is equivalent to the following condition: for each $x \in X$ the sequence of abelian groups

$$F'_x \xrightarrow{a_x} F_x \xrightarrow{b_x} F''_x$$

is exact. In this case the sequence of global sections over the space X

$$\Gamma(X, F') \longrightarrow \Gamma(X, F) \longrightarrow \Gamma(X, F'')$$

is a 0-sequence, in other words, the composition of the maps is zero, but there is no reason for this sequence to be exact. In fact, the functor $F \mapsto \Gamma(X, F)$ is a left exact functor but it is not right exact. This means that every time one has a short exact sequence

$$0 \longrightarrow F' \longrightarrow F \longrightarrow F'' \longrightarrow 0,$$

it follows that

$$0 \longrightarrow \Gamma(X, F') \longrightarrow \Gamma(X, F) \longrightarrow \Gamma(X, F'')$$

is exact but in general the last map need not surject.

A sheaf I is said to be *injective* if for each injective morphism $i : A \hookrightarrow B$, any map $f : A \to I$ extends to B. This means that there is a map $g : B \to I$ such that $g \circ i = f$. Every sheaf injects into an injective sheaf. A *complex*[1] of sheaves

$$K^{i-1} \xrightarrow{d^{i-1}} K^i \xrightarrow{d^i} K^{i+1}$$

consists of a family of sheaves K^i, indexed by $\mathbb{Z}$, and morphisms of sheaves $d^i : K^i \to K^{i+1}$ called *differentials*, such that $d^{i+1} \circ d^i = 0$. Such a complex will be denoted by $K^\bullet$. We say that $K^\bullet$ is *bounded below* if $K^i = 0$ for i sufficiently negative. Given two complexes $K^\bullet$ and $L^\bullet$, a *map of complexes* is a family of morphisms $f^i : K^i \to L^i$ such

[1] Here, these are cochain complexes: this means that the differential d^i has degree $+1$.

that the diagrams

$$\begin{array}{ccc} K^i & \xrightarrow{d^i} & K^{i+1} \\ {\scriptstyle f^i}\downarrow & & \downarrow{\scriptstyle f^{i+1}} \\ L^i & \xrightarrow{d^i} & L^{i+1} \end{array}$$

commute. The ith *cohomology sheaf* of $K^\bullet$ is the sheaf $\mathcal{H}^i(K^\bullet) = \ker d^i / \operatorname{Im} d^{i-1}$. A map of complexes $f^\bullet$ induces a map of cohomology sheaves

$$\mathcal{H}^i(f^\bullet) : \mathcal{H}^i(K^\bullet) \longrightarrow \mathcal{H}^i(L^\bullet).$$

Two maps of complexes $f^\bullet : K^\bullet \to L^\bullet$ and $g^\bullet : K^\bullet \to L^\bullet$ are said to be *homotopic* if there exists a family $h^\bullet$ of morphisms $h^i : K^i \to L^{i-1}$ such that $f^i - g^i = d^{i-1}h^i + h^{i+1}d^i$. If two morphisms of complexes are homotopic then they induce the same map on the ith cohomology sheaf.

We say that a map $f^\bullet$ is a *quasi-isomorphism* if the maps induced on the cohomology sheaves $\mathcal{H}^i(f^\bullet)$ are isomorphisms for each i. For example, this is the case when $f^\bullet$ has an inverse up to homotopy: we then say that $f^\bullet$ is a *homotopy equivalence*. An *injective resolution* of $K^\bullet$ is a pair consisting of a complex $R^\bullet$ of injective sheaves and a quasi-isomorphism $f^\bullet : K^\bullet \to R^\bullet$.

Proposition 2.1.1 *(a) It is possible to construct an injective resolution of any complex of sheaves which is bounded below.*

(b) Let $f^\bullet : K^\bullet \to L^\bullet$ be a quasi-isomorphism. For each map $\varphi^\bullet : K^\bullet \to R^\bullet$ to an injective complex, there exists a map $\psi^\bullet : L^\bullet \to R^\bullet$ for which the diagram

$$\begin{array}{ccc} K^\bullet & \xrightarrow{f^\bullet} & L^\bullet \\ & \searrow & \downarrow{\scriptstyle \psi^\bullet} \\ & & R^\bullet \end{array}$$

commutes up to homotopy. Moreover, the morphism $\psi^\bullet$ is determined uniquely up to homotopy by this condition.

Let $K^\bullet$ be a complex bounded below. We set

$$\mathbf{H}^q(X, K^\bullet) = H^q(\Gamma(X, R^\bullet)),$$

where $K^\bullet \to R^\bullet$ is an injective resolution of $K^\bullet$. By virtue of the preceding proposition, this abelian group is well defined up to isomorphism.

We call it the qth *hypercohomology group* of $K^\bullet$. Whenever the complex $K^\bullet$ consists of a single sheaf F in position 0, the resulting group is called the qth *cohomology group* of F and is denoted by $H^q(X, F)$. For $q < 0$ this group is zero and for $q = 0$ the group $H^0(X, F)$ is identified with the group $\Gamma(X, F)$ of sections of F. If $f^\bullet : K^\bullet \to K'^\bullet$ is a map of complexes then we can choose injective resolutions $R^\bullet$ and $R'^\bullet$ of $K^\bullet$ and $K'^\bullet$ to form a diagram which commutes up to homotopy,

$$\begin{array}{ccc} K^\bullet & \longrightarrow & R^\bullet \\ f^\bullet \downarrow & & \downarrow g^\bullet \\ K'^\bullet & \longrightarrow & R'^\bullet \end{array}$$

and the map $g^\bullet$ is well defined up to homotopy. It follows that we obtain a well defined map of abelian groups

$$f_* : \mathbf{H}^q(X, K^\bullet) \longrightarrow \mathbf{H}^q(X, K'^\bullet).$$

So we have an additive functor from the category of complexes bounded below to abelian groups.

A sequence of complexes

$$0 \longrightarrow K'^\bullet \longrightarrow K^\bullet \longrightarrow K''^\bullet \longrightarrow 0 \qquad (2.1)$$

is said to be *exact* if, for each integer i, it induces an exact sequence of abelian groups

$$0 \longrightarrow K'^i \longrightarrow K^i \longrightarrow K''^i \longrightarrow 0.$$

We can find injective resolutions for these three complexes which fit into a commutative diagram of exact sequences

$$\begin{array}{ccccccccc} 0 & \longrightarrow & R'^\bullet & \longrightarrow & R^\bullet & \longrightarrow & R''^\bullet & \longrightarrow & 0 \\ & & \uparrow & & \uparrow & & \uparrow & & \\ 0 & \longrightarrow & K'^\bullet & \longrightarrow & K^\bullet & \longrightarrow & K''^\bullet & \longrightarrow & 0 \end{array}$$

The fact that the abelian group R'^i is injective for each i implies that the first exact sequence induces an exact sequence of the complexes of global sections. Taking the cohomology of these complexes, we obtain a long sequence in hypercohomology:

$$\cdots \to \mathbf{H}^q(X, K'^\bullet) \to \mathbf{H}^q(X, K^\bullet) \to \mathbf{H}^q(X, K''^\bullet) \to \mathbf{H}^{q+1}(X, K'^\bullet) \to \cdots$$

The map $\delta : \mathbf{H}^q(X, K''^\bullet) \to \mathbf{H}^{q+1}(X, K'^\bullet)$ which is the only map in the preceding long exact sequence not induced by a map of complexes is called the *coboundary map*. One can check that this long exact sequence depends functorially on the initial exact sequence (2.1).

A sheaf F is said to be *acyclic* if $H^q(X, F) = 0$ for each integer $q > 0$. This hypothesis is satisfied, for example, in the following cases:

(i) The sheaf F is *flasque*: this means that the restriction map $\Gamma(X, F) \to \Gamma(U, F)$ surjects for each open subset U of X.

(ii) If the topological space X is a differentiable manifold and F is a sheaf of modules over the sheaf $\mathcal{C}_X^\infty$ of functions in C^∞ then the sheaf F is always acyclic.

Proposition 2.1.2 *(a) Each quasi-isomorphism $f : K^\bullet \to L^\bullet$ induces an isomorphism $f_* : \mathbf{H}^q(X, K^\bullet) \to \mathbf{H}^q(X, L^\bullet)$ in hypercohomology. In particular, if $K^\bullet$ is an exact complex then its hypercohomology vanishes.*

(b) If K^p is acyclic for each p then the canonical map

$$H^q(\Gamma(X, K^\bullet)) \to \mathbf{H}^q(X, K^\bullet)$$

is an isomorphism for each q.

Examples.

(1) The last statement constitutes a formal version of the classical theorem of de Rham. Consider the complex of sheaves $\mathcal{A}^\bullet$ over a smooth manifold called the *de Rham complex* consisting of complex differential forms. Given the above statement, we have an isomorphism

$$H^q(\Gamma(X, \mathcal{A}^\bullet)) \cong \mathbf{H}^q(X, \mathcal{A}^\bullet).$$

For the other part, the Poincaré Lemma says that, over an open convex set of $\mathbb{R}^n$, the complex of differential forms is exact in all degrees above zero, this implies that the map $\mathbb{C} \to \mathcal{A}^\bullet$ of complexes, where $\mathbb{C}$ denotes the complex consisting of the constant sheaf $\mathbb{C}$ in degree 0, is a quasi-isomorphism. It follows from the first part of the statement that one has a canonical isomorphism

$$\mathbf{H}^q(X, \mathbb{C}) \cong H^q(\Gamma(X, \mathcal{A}^\bullet)).$$

(2) Over a smooth algebraic variety we can consider the sheaves Ω_X^q of regular differentiable forms: these are given as the sheaf of regular local sections of the vector bundle $\Lambda^q T^*(X)$. We can construct a derivation $d : \Omega_X^q \to \Omega_X^{q+1}$ characterized by the following properties:

(a) If α and β are regular differentiable forms of degrees p and q respectively then the Leibniz formula holds:

$$d(\alpha \wedge \beta) = d\alpha \wedge \beta + (-1)^p \alpha \wedge d\beta.$$

(b) If f is a regular function on an open subset of X then df is the differential of f.

In this way we obtain a complex of sheaves of differential forms $\Omega_X^\bullet$. The hypercohomology $\mathbf{H}^q(X, \Omega_X^\bullet)$ is called the *de Rham cohomology* of X. This cohomology vanishes in degree $q > 2 \dim X$ as can be easily seen after the following result of Grothendieck:

Proposition 2.1.3 *Let X be a Noetherian topological space of dimension n and let F be a sheaf of abelian groups on X. Then $H^q(X, F) = 0$ for $q > n$.*

In fact, it is possible to show that the de Rham cohomology defined here uniquely in algebraic terms can be identified with the de Rham cohomology for the underlying C^∞ manifolds as described above.

(3) If X is irreducible then every constant sheaf G defined over the open set U by the locally constant functions on U to a fixed abelian group G is flasque. From this we can see that such a sheaf is forced to be acyclic: $H^q(X, G) = 0$ for $q > 0$.

Suppose that the variety X is still irreducible. Consider the exact sequence of sheaves of abelian groups on X

$$0 \longrightarrow \mathcal{O}_X^* \longrightarrow \mathcal{K}^* \longrightarrow \underline{\mathrm{Div}}_X \longrightarrow 0,$$

where $\mathcal{K}^*$ is the sheaf of non-zero constant rational functions and $\underline{\mathrm{Div}}_X$ is the sheaf of (Cartier) divisors of X. The preceding exact sequence can be viewed as a flasque resolution of $\mathcal{O}_X^*$. We can then deduce that $H^q(X, \mathcal{O}_X^*) = 0$ for $q \geq 2$ and the group $H^1(X, \mathcal{O}_X^*)$ can be identified with the quotient of the group $\Gamma(X, \underline{\mathrm{Div}}_X)$ given by Cartier divisors modulo linear equivalence. One can verify that this group is also the Picard group of isomorphism classes of invertible vector bundles (in other words, of rank one); the group structure is given by the tensor product.

The method described to define hypercohomology and cohomology is, in fact, a particular case of the construction of right hyperderived functors $\mathbb{R}^q\mathbf{F}$ and derived functors $R^q\mathbf{F}$ of an additive functor $\mathbf{F} : \mathcal{A} \to \mathcal{B}$ from one abelian category $\mathcal{A}$ to another, $\mathcal{B}$. The only condition which one must impose is that each object of $\mathcal{A}$ must inject into an injective object. In the same way we can define, for each additive functor $\mathbf{F}$, the left hyperderived (and left derived) functor on the category of chain complexes $K_\bullet$ (in other words, complexes for which the differential $d_i : K_i \to K_{i-1}$ has degree -1) which vanish when the degree is sufficiently negative by using projective resolutions $P_\bullet \to K_\bullet$; this makes sense if all objects are quotients of projective objects. These functors are particularly interesting for right exact functors.

2.2 Coherent algebraic sheaves

Let X be an algebraic variety with structure sheaf $\mathcal{O}_X$. A sheaf of modules over $\mathcal{O}_X$ is called an *algebraic* sheaf. The category of algebraic sheaves is obviously abelian and any algebraic sheaf injects into an injective algebraic sheaf so that one can define the notion of injective resolution of a complex $K^\bullet$ of algebraic sheaves bounded below and hence its hypercohomology $\mathbf{H}^q(X, K^\bullet)$: these are modules over the algebra $\mathcal{O}(X)$ of global regular functions. In fact, injective sheaves in this category are always flasque so that, by Proposition 2.1.2, the underlying abelian group structure is none other than that of the hypercohomology of the underlying sheaf of abelian groups.

An algebraic coherent sheaf is said to be *coherent* if

(a) for each point $x \in X$ there exist an open neighbourhood U of x and a surjective map over U

$$\mathcal{O}_X^p|_U \longrightarrow F|_U$$

(we say that such a sheaf F is *of finite type*),

(b) for each map

$$\mathcal{O}_X^q|_U \xrightarrow{f} F|_U,$$

the sheaf of relations of F given by $\ker f$ is also of finite type.

The coherent algebraic sheaves on X form an abelian category $\mathrm{Coh}(X)$. The structure sheaf is coherent so that one can characterize coherent sheaves by the following property: for each $x \in X$, there exist an open neighbourhood U of x, locally-free sheaves $\mathcal{L}_1$ and $\mathcal{L}_0$, and an exact sequence of sheaves over the open set U

$$\mathcal{L}_1 \longrightarrow \mathcal{L}_0 \longrightarrow F|_U \longrightarrow 0.$$

Theorem 2.2.1 (Serre) *Let X be an affine algebraic variety and F a coherent algebraic sheaf on X.*

(i) One has $H^q(X, F) = 0$ *for each* $q > 0$.

(ii) The sheaf F *is generated by a finite number of sections; in other words, there exists a surjective map*

$$\mathcal{O}_X^p \longrightarrow F.$$

(iii) The functor $\mathrm{Coh}(X) \to \mathrm{Mod}(\mathcal{O}(X))$ *from the category of coherent algebraic sheaves on* X *to that of finitely generated modules over the algebra* $A = \mathcal{O}(X)$ *is an equivalence of categories. This functor has as its quasi-inverse the functor which associates to the finitely generated* A*-module* M *the coherent algebraic sheaf* $F = M \otimes_A \mathcal{O}_X$.

In the equivalence of categories above, the locally-free sheaves correspond to the projective finitely generated modules.

Suppose that X is a smooth affine algebraic variety. The first part of the theorem shows that the sheaves of differential forms Ω_X^p, which are coherent because they are locally-free, are acyclic. Thus, the de Rham cohomology groups defined by $H^q_{\mathrm{DR}}(X) = \mathbf{H}^q(X, \Omega_X^\bullet)$ are given by the cohomology of the complex of global regular differential forms:

$$H^q_{\mathrm{DR}}(X) = H^q(\Gamma(X, \Omega_X^\bullet)).$$

2.3 Projective geometry

In this section we assume that X is a projective algebraic variety together with an embedding $X \hookrightarrow \mathbf{P}_N(\mathbb{C})$ into projective space. We let $\mathcal{O}_X(n)$ denote the bundle induced by $\mathcal{O}_{\mathbf{P}_N(\mathbb{C})}(n)$, and we use the same notation for its sheaf of sections. For each algebraic sheaf X one sets

$$F(n) = F \otimes_{\mathcal{O}_X} \mathcal{O}_X(n).$$

Theorem 2.3.1 (Finiteness) *Let* F *be a coherent algebraic sheaf over a projective variety* X. *Then the cohomology groups* $H^q(X, F)$ *are* $\mathbb{C}$*-vector spaces of finite dimension.*

We set $h^q(X, F) = \dim H^q(X, F)$ and, by Grothendieck's Theorem (Proposition 2.1.3), we know that $h^q(X, F) = 0$ for $q > \dim X$. This justifies the formula

$$\chi(X, F) = \sum_q (-1)^q h^q(X, F).$$

This number is called the *Euler-Poincaré characteristic* of F. If we have a short exact sequence

$$0 \longrightarrow F' \longrightarrow F \longrightarrow F'' \longrightarrow 0$$

of coherent algebraic sheaves on X then we have $\chi(X,F) = \chi(X,F') + \chi(X,F'')$.

Theorem 2.3.2 *Let F be a coherent algebraic sheaf on X. For n large enough*

(A) the sheaf $F(n)$ is generated by its sections and
(B) we have $H^q(X,F(n)) = 0$ for $q > 0$.

Example.
Let X be a smooth projective algebraic curve. Then every coherent algebraic sheaf F is the cokernel of an injective map of locally-free sheaves, in other words, there exist locally-free sheaves $\mathcal{L}_1$ and $\mathcal{L}_0$ on X and an exact sequence

$$0 \longrightarrow \mathcal{L}_1 \longrightarrow \mathcal{L}_0 \longrightarrow F \longrightarrow 0.$$

Indeed, let the integer n be given as in the theorem. If we set $k = h^0(X, F(n))$ and $\mathcal{L}_0 = \mathcal{O}_X^k(-n)$ then, by (A), there is a surjective map

$$\varphi : \mathcal{L}_0 \longrightarrow F.$$

Now, it suffices to show that the kernel $\mathcal{L}_1 = \ker\varphi$ is locally-free. Since this is a coherent sheaf, it suffices to prove that the module $\mathcal{L}_{1,x}$ is a free module over $\mathcal{O}_{X,x}$ for each $x \in X$. We can now use the fact that the finitely generated torsion-free modules over a principal ring are free, or the following lemma:

Lemma 2.3.3 *Let A be a local ring with maximal ideal $\mathfrak{m}$, $k = A/\mathfrak{m}$ its residue field and M a finitely generated module over A. Then M is free if and only if*

$$\mathrm{Tor}_1^A(M,k) = 0.$$

The functors $M \mapsto \mathrm{Tor}_1^A(M,k)$ are the left derived functors of the (right exact) functor $M \mapsto M \otimes_A k$, so they are computed from a projective resolution of M or of k. We apply the lemma to the case when A is the local ring $\mathcal{O}_{X,x}$ of a smooth curve at a point x so that the maximal ideal $\mathfrak{m}$ is generated by a single element and we have a free resolution

$$0 \longrightarrow A \longrightarrow A \longrightarrow k \longrightarrow 0.$$

Then, for each A-module M, we have $\mathrm{Tor}_p^A(M,k) = 0$ for $p > 1$. From the exact sequence above it follows that

$$0 = \mathrm{Tor}_2^A(F_x, k) \cong \mathrm{Tor}_1^A(\mathcal{L}_{1,x}, k).$$

Hence, $\mathcal{L}_{1,x}$ is a free module by the lemma.

Proof We choose a basis $(\bar{e}_1, \ldots, \bar{e}_r)$ of the k-vector space $M/\mathfrak{m}M = M \otimes_A k$. These elements lift to elements $(e_1, \ldots, e_r)$ of M which, by Nakayama's Lemma, generate M. This system of generators then give rise to a surjective map $A^r \to M$ of modules and we let N be the kernel. Thc long exact sequence associated to the sequence

$$0 \longrightarrow N \longrightarrow A^r \longrightarrow M \longrightarrow 0$$

after applying the functor $- \otimes_A k$ is

$$0 \longrightarrow \mathrm{Tor}_1^A(M,k) \longrightarrow N \otimes_A k \longrightarrow k^r \longrightarrow M \otimes_A k$$

and the hypothesis shows that $N \otimes_A k = 0$; then $N = 0$ by Nakayama's Lemma. The statement now follows. □

2.4 Algebraic geometry and analytic geometry

To be given an algebraic variety means that we have a ringed space $(X, \mathcal{O}_X)$ for which each point has an open neighbourhood U which is locally isomorphic to an affine sub-variety of $\mathbb{C}^k$ defined by an ideal in the algebra of polynomials on $\mathbb{C}^k$ and these must satisfy the usual separation and finiteness conditions. We can view this local model as an analytic space and the transition functions are also analytic. Hence, we can associate to X an analytic space X^{an} with structure sheaf $\mathcal{O}_X^{\mathrm{an}}$. There is a morphism of ringed spaces

$$j : (X^{\mathrm{an}}, \mathcal{O}_X^{\mathrm{an}}) \longrightarrow (X, \mathcal{O}_X)$$

which means that every Zariski open set U of X is also open in the usual topology, that is, in the topology of X^{an}, and there is a map of algebras $j^* : \Gamma(U, \mathcal{O}_X) \to \Gamma(U, \mathcal{O}_X^{\mathrm{an}})$. If we are given a sheaf F on X then we can define an inverse image which will be denoted by F^{an}: this is the associated analytic sheaf which has as its module of germs of sections in a neighbourhood of x

$$F_x^{\mathrm{an}} = F_x \otimes_{\mathcal{O}_{X,x}} \mathcal{O}_{X,x}^{\mathrm{an}}.$$

An essential result, which was proved by Serre, is the following:

Theorem 2.4.1 *For each point $x \in X$, the local ring $B = \mathcal{O}^{\text{an}}_{X,x}$ is a flat module over $A = \mathcal{O}_{X,x}$.*

This implies that the functor defined on the category of A-modules by $M \mapsto M \otimes_A B$ sends exact sequences to exact sequences. Given a complex of algebraic sheaves $K^\bullet$ and an injective resolution $K^\bullet \to L^\bullet$ this functor furnishes us with a resolution $K^{\bullet,\text{an}} \to R^{\bullet,\text{an}}$ of the complex of associated analytic sheaves. It follows that we obtain a map in hypercohomology:

$$\mathbf{H}^q(X, K^\bullet) \longrightarrow \mathbf{H}^q(X, K^{\bullet,\text{an}}).$$

Theorem 2.4.2 *If X is a projective variety then, for all complexes $K^\bullet$ of coherent algebraic sheaves, the canonical map*

$$\mathbf{H}^q(X, K^\bullet) \longrightarrow \mathbf{H}^q(X, K^{\bullet,\text{an}})$$

is an isomorphism.

Example.
Suppose that the algebraic variety X is projective and smooth and consider the locally-free sheaf Ω^p_X of regular differential forms of degree p and the sheaf $\Omega^p_{X^{\text{an}}}$ of holomorphic forms on the smooth variety X^{an}. Then the map

$$H^q(X, \Omega^p_X) \longrightarrow H^q(X^{\text{an}}, \Omega^p_{X^{\text{an}}})$$

is an isomorphism for each pair (p, q). We deduce that the de Rham cohomology $H^q_{\text{DR}}(X)$ can be identified with the hypercohomology

$$\mathbf{H}^q(X^{\text{an}}, \Omega^\bullet_{X^{\text{an}}})$$

of the complex $\Omega^\bullet_{X^{\text{an}}}$ of holomorphic forms on X^{an}. Now, the canonical map $\Omega^\bullet_{X^{\text{an}}} \to \mathcal{A}^\bullet_{X^{\text{an}}}$ which takes values in the de Rham complex of C^∞ differential forms is a quasi-isomorphism as one can see by working over small contractible open sets. It follows that the de Rham cohomology coincides with the usual cohomology.

Corollary 2.4.3 *Let b_q be the q-th Betti number of X^{an}. Then we have*

$$\dim H^q_{\text{DR}}(X) = b_q.$$

So this number is an invariant which only depends on the usual topology of X. The same can also be shown if X is not projective.

2.5 Serre duality

Consider a connected smooth projective variety of dimension n. Then, it follows from the preceding section, we have $H^{2n}_{\mathrm{DR}}(X) \cong \mathbb{C}$; the isomorphism is canonical and is obtained via integration of differential forms. We set $\omega_X = \Omega^n_X$: this is the *canonical sheaf* of X. There is an obvious canonical map $H^n(X, \omega_X) \to H^{2n}_{\mathrm{DR}}(X) = \mathbb{C}$. Consider a locally-free sheaf F over X, an injective resolution $R^\bullet$ of F and let F^* denote the dual of F. Let $I^\bullet$ be an injective resolution of ω_X. Then $F^* \otimes_{\mathcal{O}_X} I^\bullet$ is still an injective resolution of $F^* \otimes_{\mathcal{O}_X} \omega_X$ and the cohomology vector space $H^q(\Gamma(X, F^* \otimes_{\mathcal{O}_X} I^\bullet))$ can be identified with the vector space of homotopy classes of maps of complexes from the complex consisting of just F to the shifted complex $I^\bullet(q)$ of $I^\bullet$ defined by $I^\bullet(q)^i = I^{q+i}$. It follows that such a homotopy class α induces a map $R^\bullet \to I^\bullet(q)$ which is also defined up to homotopy. This then defines a linear map $\sigma(\alpha) : H^{n-q}(X, F) \to H^n(X, \omega_X) \to \mathbb{C}$ which depends linearly on α. So we have a linear map

$$\sigma : H^q(X, F^* \otimes \omega_X) \longrightarrow H^{n-q}(X, F)^*.$$

Theorem 2.5.1 (Serre) *The linear map σ above is an isomorphism for each integer q.*

Note that if we only have a coherent algebraic sheaf instead of a locally-free sheaf then we can introduce the derived functors $\mathrm{Ext}^q_{\mathcal{O}_X}(F, -)$ of the functor $\mathrm{Hom}_{\mathcal{O}_X}(F, -)$ which is left exact and we also obtain a $\mathbb{C}$-linear map

$$\sigma : \mathrm{Ext}^q_{\mathcal{O}_X}(F, \omega_X) \longrightarrow H^{n-q}(X, F)^*.$$

The Grothendieck Duality Theorem then tells us that this linear map is still an isomorphism.

Theorem 2.5.2 *Let F be a locally-free sheaf on X. Then we have*

$$\chi(X, F) = (-1)^n \chi(F^* \otimes_{\mathcal{O}_X} \omega_X).$$

2.6 The Riemann-Roch formula for curves

The notion of rank

Let F be a coherent algebraic sheaf on an algebraic variety X.

Lemma 2.6.1 *If the variety X is integral then there exists an open, everywhere dense subset U of X on which $F|_U$ is locally-free.*

Proof In fact, it suffices to find a non-empty open subset V of X on which the sheaf F admits a presentation

$$\mathcal{L}_1 \xrightarrow{f} \mathcal{L}_0 \longrightarrow F|_V \longrightarrow 0$$

by locally-free sheaves, that is, such that this sequence is exact. The rank of the matrix given by f is a lower semi-continuous function. Consequently, the points where the rank is maximal form an open set U over which the rank is constant. Then the sheaf $\operatorname{coker} f$ is locally-free over this open set by Proposition 1.7.2. □

Note that this statement will fail over a non-reduced variety: to convince ourselves of this it suffices to consider the structure sheaf of a closed point in the variety defined by a double point which consists of a point equipped with the ring of dual numbers $\mathbb{C}[\epsilon]/(\epsilon^2)$.

We call the rank of the locally-free sheaf the *rank* of F; it is obviously independent of the choice of open set U. We let $\operatorname{rk}(F)$ denote the rank of F. If $0 \to F' \to F \to F'' \to 0$ is an exact sequence of coherent algebraic sheaves on X then we have $\operatorname{rk}(F) = \operatorname{rk}(F') + \operatorname{rk}(F'')$. If the support of F is different from X then the rank is clearly 0. Then we say that the sheaf F is a *torsion* sheaf: this property is, in fact, equivalent to the fact that for each point $x \in X$, the module of the germs of sections F_x is, if it is not zero, a torsion $\mathcal{O}_{X,x}$-module.

Let X be a smooth connected projective curve. The genus of X is the number $g = h^1(\mathcal{O}_X)$. Then we have $\chi(\mathcal{O}_X) = 1 - g$.

Proposition 2.6.2 *Let $b_1 = \dim H^1_{\mathrm{DR}}(X)$ be the first Betti number of X. We have the relation $b_1 = 2g$. In particular, the genus of X is a topological invariant.*

Proof Consider the following exact sequences of complexes on X:

$$\begin{array}{ccccccccc}
 & & & & 0 & & 0 & & \\
 & & & & \downarrow & & \downarrow & & \\
 & & 0 & \longrightarrow & \mathcal{O}_X & \longrightarrow & \mathcal{O}_X & \longrightarrow & 0 \\
 & & \downarrow & & \downarrow & & \downarrow & & \\
0 & \longrightarrow & \Omega^1_X & \longrightarrow & \Omega^1_X & \longrightarrow & 0 & & \\
 & & \downarrow & & \downarrow & & & & \\
 & & 0 & & 0 & & & &
\end{array}$$

It gives rise to an exact sequence

$$0 \to H^0(\Omega^1_X) \to H^1_{\mathrm{DR}}(X) \to H^1(\mathcal{O}_X) \to 0.$$

By Serre duality, $h^0(\Omega^1_X) = h^1(\mathcal{O}_X) = g$ and so $b_1 = 2g$. □

Example.
Let X be a smooth curve of degree n in the projective plane $\mathbf{P}_2$. Then X is connected and $g = \frac{1}{2}(n-1)(n-2)$. Indeed, we have the exact sequence of sheaves on the projective plane $\mathbf{P}_2$

$$0 \longrightarrow \mathcal{O}_{\mathbf{P}_2}(-n) \longrightarrow \mathcal{O}_{\mathbf{P}_2} \longrightarrow \mathcal{O}_X \longrightarrow 0$$

and it follows that $\chi(\mathcal{O}_X) = \chi(\mathcal{O}_{\mathbf{P}_2}) - \chi(\mathcal{O}_{\mathbf{P}_2}(-n))$. The formula is the result of the following facts: the canonical sheaf of the projective plane is given by $\omega_{\mathbf{P}_2} = \mathcal{O}_{\mathbf{P}_2}(-3)$ and the cohomology groups $H^q(\mathcal{O}_{\mathbf{P}_2}(k))$ vanish if $q > 0$ and $k > -3$. From the Serre Duality Theorem we have $\chi(\mathcal{O}_{\mathbf{P}_2}(-n)) = \chi(\mathcal{O}_{\mathbf{P}_2}(n-3))$; on the other hand, $\chi(\mathcal{O}_{\mathbf{P}_2}) = 1$. Then we obtain $\chi(\mathcal{O}_X) = 1 - \frac{1}{2}(n-1)(n-2)$ from which we can recover the formula for the genus.

The degree

Given a coherent algebraic sheaf F of rank r on a smooth curve X we define the *degree* $d = d(F)$ by the formula

$$d(F) = \chi(F) - r\chi(\mathcal{O}_X).$$

If $0 \to F' \to F \to F'' \to 0$ is an exact sequence of coherent algebraic sheaves on X then we have the addition formula $d(F) = d(F') + d(F'')$ which follows from the additivity of the rank and the Euler-Poincaré characteristic.

Let $D = \sum \nu_i a_i$ be an effective divisor on a curve X. The ideal I_D of $\mathcal{O}_X$ of the locally regular functions which annihilate a_i with multiplicity at least ν_i is locally-free of rank 1. Consider the invertible bundle L_D defined by $I_D = \mathcal{O}(L_D^*)$. The inclusion map

$$\mathcal{O}(L_D^*) \hookrightarrow \mathcal{O}_X$$

defines a section of L_D whose scheme of zeros is D. Consider the group $\mathrm{Div}(X)$ of divisors of X and the Picard group $\mathrm{Pic}(X)$ of invertible bundles. The map $D \mapsto L_D$ extends to a surjective map $\lambda : \mathrm{Div}(X) \to \mathrm{Pic}(X)$: indeed, if L is an invertible bundle on X then, for n large enough, $L(n)$ has a section which, by Serre's Theorem A, does not vanish identically. The kernel of this map consists of the sub-group of principal divisors; that is, the divisors of non-zero rational functions. The equivalence relation $D \sim D'$ for $D - D' \in \ker\lambda$ is called *linear equivalence.*

If $D = \sum \nu_i a_i$ is some divisor on X then the number $\deg(D) = \sum_i \nu_i$ is called the *degree* of D.

Theorem 2.6.3 *Let D be a divisor of X. We have $d(L_D) = \deg(D)$.*

Proof Suppose first that $D = \sum \nu_i a_i$ is an effective divisor. Regarded as a sub-scheme of dimension 0 of X we have

$$\mathcal{O}_D = \bigoplus(\mathcal{O}_X/\mathfrak{m}_{a_i}^{\nu_i}),$$

where $\mathfrak{m}_{a_i}$ is the ideal of $\mathcal{O}_X$ given by the point a_i. It follows that $d(\mathcal{O}_D) = \deg(D)$. The bundle L_D is trivial in a neighbourhood of D and we therefore have the exact sequence $0 \to \mathcal{O}_X \to \mathcal{O}_X(L_D) \to \mathcal{O}_D \to 0$ from which we get $d(L_D) = \deg(\mathcal{O}_D) = \deg(D)$. Now consider any divisor D which we write as $D = B - A$, where A and B are effective divisors. The bundle L associated to D is then $L = L_B \otimes L_A^*$. We also have the following exact sequence of sheaves on X associated to the canonical section of L_A:

$$0 \longrightarrow L \longrightarrow L_B \longrightarrow \mathcal{O}_A \longrightarrow 0.$$

It follows that $d(L) = d(L_B) - d(A)$ and, from the result which we have already proved for the effective divisor B, we obtain $d(L) = d(B) - d(A) = \deg(D)$. □

Corollary 2.6.4 *The map $L \mapsto d(L) : \mathrm{Pic}(X) \to \mathbb{Z}$ is a group homomorphism.*

Corollary 2.6.5 *If an invertible bundle L on X has a section which does not vanish identically then we have $d(L) \geq 0$. The degree of L is the number of zeros of a non-zero section of L counted with multiplicity; it is independent of the choice of section.*

Grothendieck groups

Consider the group $K(X)$ (respectively, $K_0(X)$) which is the quotient of the free group generated by the coherent algebraic sheaves (respectively, the locally-free sheaves) on the curve X by the sub-group generated by the elements $F - F' - F''$ for each exact sequence $0 \to F' \to F \to F'' \to 0$. There is a canonical map $i : K_0(X) \to K(X)$. We shall prove the following proposition only for a smooth curve but, in fact, the result holds for all smooth varieties.

Proposition 2.6.6 *The canonical map $i : K_0(X) \to K(X)$ is an isomorphism.*

Proof We shall construct an inverse j of the map i. Consider a coherent algebraic sheaf F and a resolution $0 \to \mathcal{L}_1 \to \mathcal{L}_0 \to F \to 0$ by locally-free sheaves. We set

$$j(F) = \mathcal{L}_0 - \mathcal{L}_1$$

in $K_0(X)$. We need to prove that this element is independent of the resolution we have chosen. Let $0 \to \mathcal{L}'_1 \to \mathcal{L}'_0 \to F \to 0$ be another resolution. Consider the canonical map $\mathcal{L}_0 \oplus \mathcal{L}'_0 \to F$. This map is surjective and its kernel N fits into an exact sequence $0 \to \mathcal{L}_1 \to N \to \mathcal{L}'_0 \to 0$. Then, in $K_0(X)$, we have $N = \mathcal{L}_1 + \mathcal{L}'_0$ and, by symmetry, we also have $N = \mathcal{L}'_1 + \mathcal{L}_0$. It follows that, in $K_0(X)$,

$$\mathcal{L}_1 + \mathcal{L}'_0 = \mathcal{L}'_1 + \mathcal{L}_0$$

from which we have $\mathcal{L}_0 - \mathcal{L}_1 = \mathcal{L}'_0 - \mathcal{L}'_1$.

It remains to verify that, whenever we have an exact sequence $0 \to F' \to F \to F'' \to 0$, we have $j(F) = j(F') + j(F'')$. We choose surjective maps η', η'':

$$\begin{array}{ccccccccc} & & \mathcal{L}'_0 & & & & \mathcal{L}''_0 & & \\ & & {\scriptstyle \eta'}\downarrow & & & & \downarrow{\scriptstyle \eta''} & & \\ 0 & \longrightarrow & F' & \longrightarrow & F & \longrightarrow & F'' & \longrightarrow & 0 \end{array}$$

with $\mathcal{L}'_0$ a locally-free sheaf and $\mathcal{L}''_0 = \mathcal{O}_X(-m)^k$, with m sufficiently large so that the map η'' lifts to a map $\mathcal{L}''_0 \to F$. The choice of such a lift gives us a canonical map $\mathcal{L}_0 = \mathcal{L}'_0 \oplus \mathcal{L}''_0 \to F$. Then we have a commutative diagram

$$\begin{array}{ccccccccc} 0 & \longrightarrow & \mathcal{L}'_0 & \longrightarrow & \mathcal{L}_0 & \longrightarrow & \mathcal{L}''_0 & \longrightarrow & 0 \\ & & {\scriptstyle \eta'}\downarrow & & {\scriptstyle \eta}\downarrow & & \downarrow{\scriptstyle \eta''} & & \\ 0 & \longrightarrow & F' & \longrightarrow & F & \longrightarrow & F'' & \longrightarrow & 0 \end{array}$$

in which the vertical maps are surjections. We let $\mathcal{L}'_1$, $\mathcal{L}_1$ and $\mathcal{L}''_1$ be the respective kernels of the vertical maps. These coherent sheaves are locally-free and so we have a short exact sequence $0 \to \mathcal{L}_1' \to \mathcal{L}_1 \to \mathcal{L}_1'' \to 0$ which gives the formula $\mathcal{L}_1 = \mathcal{L}_1' + \mathcal{L}_1''$ in $K_0(X)$. It follows that $j(F) = j(F') + j(F'')$ which is the formula we required. □

Corollary 2.6.7 *The Grothendieck group $K(X)$ admits the structure of a ring (with unity) induced by the tensor product of locally-free sheaves.*

For each vector bundle F of rank r on X we set

$$\det(F) = \Lambda^r(F)$$

in $\mathrm{Pic}(X)$. This is the *determinant bundle.*

Corollary 2.6.8 *The map $F \mapsto \det(F)$ induces a homomorphism of abelian groups $K(X) \to \mathrm{Pic}(X)$.*

This allows us to speak of the determinant bundle of a coherent algebraic sheaf on X: it is well defined up to isomorphism. For example, if D is a sub-scheme of X of dimension 0 and $\mathcal{O}_D$ its structure sheaf then we have $\det(\mathcal{O}_D) = L_D$.

Theorem 2.6.9 *Let F be a coherent algebraic sheaf on X. Then we have $d(F) = d(\det(F))$.*

Proof It is clear that the map $F \mapsto d(F)$ extends to $K(X)$ to give a homomorphism of abelian groups $d : K(X) \to \mathbb{Z}$. The two sides of the equation are then group homomorphisms so that it suffices to verify the equation for a system of generators of $K(X)$. This is going to follow from the following description of $K(X)$. □

Lemma 2.6.10 *Let $A(X)$ be the sub-group of $K(X)$ generated by the collection of the structure sheaves of points. Then we have*

$$K(X) = A(X) \oplus \mathbb{Z}.$$

Proof It is evident that $A(X) \cap \mathbb{Z} = 0$ by looking at the ranks. So it suffices to prove that $A(X) + \mathbb{Z} = K(X)$. Any element $\alpha \in K(X)$ can be written as $\alpha = F - G$, where F and G are vector bundles. Then it suffices to show that each vector bundle is contained in this sum. If the bundle has rank r and is generated by its sections we can find a map $f : \mathcal{O}^r \to F$ which is an isomorphism over a fixed point. Then we have an exact sequence of sheaves

$$0 \longrightarrow \mathcal{O}^r \longrightarrow F \longrightarrow \Theta \longrightarrow 0,$$

where Θ is a sheaf on the sub-scheme given by the section $\det(f)$ of the bundle $\det(F)$. This sub-scheme has dimension 0 and the structure theorem of torsion modules of finite type over principal rings allows us to write Θ as a finite direct sum of sheaves $\Theta_{\nu,x}$ of type

$$\Theta_{\nu,x} = \mathcal{O}_X/\mathfrak{m}_x^\nu,$$

where ν is an integer greater than or equal to 1 and x is a point of X. In the Grothendieck group we clearly have $\Theta_{\nu,x} = \nu\mathbb{C}_x$, where $\mathbb{C}_x$ is the structure sheaf of the point x. Thus, $\Theta_{\nu,x}$ and hence Θ is contained in $A(X)$.

It follows that, for each bundle F and each integer n large enough, $F(n)$ and $\mathcal{O}(n)$ are contained in $A(X)+\mathbb{Z}$. On the other hand, it is clear that every element of $A(X)$ has square zero in $K(X)$: therefore, $\mathcal{O}(-n)$ is contained in this sub-group and this sub-group is, in fact, a sub-ring. It follows that $K(X) = A(X) \oplus \mathbb{Z}$ which proves the lemma. □

Proof of Theorem 2.6.9 It suffices to verify the formula for the structure sheaf $\mathcal{O}_D$ of a dimension 0 sub-scheme of X. In this case, we have $\det(\mathcal{O}_D) = L_D$ and, by Theorem 2.6.3, we obtain $d(\det(\mathcal{O}_D)) = d(L_D) = \deg(D)$. Evidently, $d(\mathcal{O}_D) = \deg(D)$, which proves the theorem. □

Remark.
It follows from Lemma 2.6.10 that the abelian group homomorphism $K(X) \to \mathrm{Pic}(X) \oplus \mathbb{Z}$ defined by $F \mapsto (\det(F), \mathrm{rk}(F))$ is an isomorphism.

2.7 Exercises

Let X be a smooth projective curve.

2.1 Show that every torsion-free coherent algebraic sheaf F is locally free. (Recall the structure of the modules of finite type over principal rings.)

2.2 Show that the dual $F^* = \mathcal{H}om_{\mathcal{O}_X}(F, \mathcal{O}_X)$ of a coherent algebraic sheaf F on X is locally-free.

2.3 Let F be a coherent algebaric sheaf on X and T be the kernel of the canonical map $F \to F^{**}$. Show that T has rank 0 and that F is isomorphic to the direct sum of T and a locally-free sheaf.

2.4 Let $\alpha \in K(X)$ (respectively, $\alpha' \in K(X)$) be an element of rank r (respectively rank r') and of degree d (respectively d'). Show that $\alpha \cdot \alpha'$ has rank rr' and degree $rd' + r'd$.

2.5 Determine the signature of the quadratic form $K(X) \to \mathbb{Z}$ defined by $\alpha \mapsto \chi(\alpha^2)$.

2.6 Let E be a locally-free sheaf on X. Consider the formal series with coefficients in $K(X)$ given by

$$s(E) = \sum_{i \geq 0} t^i S^i(E), \qquad \lambda(E) = \sum_{i \geq 0} t^i \Lambda^i(E).$$

(a) Show that, for each exact sequence of locally-free sheaves

$$0 \longrightarrow E' \longrightarrow E \longrightarrow E'' \longrightarrow 0,$$

we have $s(E) = s(E')s(E'')$ and $\lambda(E) = \lambda(E')\lambda(E'')$. Deduce that we can define $s(\alpha)$ and $\lambda(\alpha)$ for any $\alpha \in K(X)$. We set

$$s(\alpha) = \sum_{i \geq 0} t^i s^i(\alpha), \qquad \lambda(\alpha) = \sum_{i \geq 0} t^i \lambda^i(\alpha).$$

(b) Calculate the rank and the degree of $s^i(\alpha)$ and of $\lambda^i(\alpha)$.

3

Topological Classification

Taking the definition we have given of algebraic vector bundles as a model, we can define the notion of topological vector bundle on any topological space. The classification of complex topological vector bundles of rank r on the sphere S_{n+1} leads to a description of the group of homotopy classes of continuous maps $S_n \to \mathrm{GL}(r, \mathbb{C})$. In the first section of this chapter we explain how the computation of this group reduces to the computation of the homotopy groups of spheres; this computation can be done for $n \leq 3$. It is used as the starting point in the classification of complex topological vector bundles over CW-complexes of dimension at most 4 in terms of rank and Chern classes (Wu's Theorem [61]). For more details about homotopy groups of the classical groups and the connection with vector bundles, one can consult the book of M.F. Atiyah [2] or that of D. Husemoller [33].

We will only be concerned with the case of compact Riemann surfaces and then we only need to know the Poincaré groups of $\mathrm{GL}(r, \mathbb{C})$. So the complex topological vector bundles of rank r on such a Riemann surface are classified by a number d called the degree. Of course for algebraic vector bundles this number coincides with the degree which we introduced in chapter 2. For topological vector bundles of rank 1, the definition of degree to be given in section 3.5 in a certain sense counts the number of zeros of a section which has only finitely many zeros: the essential point is to show that this number does not depend on the section. As we shall see in section 3.6, this result can be obtained by introducing the notion of Chern classes.

3.1 Homotopy groups

Let (X, a) be a topological space with a fixed base point. We let I denote the interval $[0, 1]$ and $\pi_n(X, a)$ the set of homotopy classes of maps from $(I^n, \partial I^n)$ to (X, a); it is given a group structure by setting

$$\alpha.\beta(t_1, \ldots, t_n) = \begin{cases} \alpha(t_1, \ldots, 2t_n) & \text{if } 0 \le t_n \le \frac{1}{2}, \\ \beta(t_1, \ldots, 2t_n - 1) & \text{if } \frac{1}{2} \le t_n \le 1 \end{cases}$$

for two continuous functions α and $\beta : (I^n, \partial I^n) \to (X, a)$. Let $\Omega(X)$ denote the space of loops in X based at a. We can also write $\pi_n(X, a) = \pi_1(\Omega(\cdots \Omega(X) \cdots))$ which implies that $\pi_n(X, a)$ is an abelian group for $n \ge 2$. A continuous function $f : (X, a) \to (Y, a)$ induces a group homomorphism $f_* : \pi_n(X, a) \to \pi_n(Y, b)$ which only depends on the homotopy class of f.

Properties

(1) We have

$$\begin{aligned} \pi_i(S_n) &= 0 \qquad \text{for } i < n, \\ \pi_n(S_n) &= \mathbb{Z}, \\ \pi_n(S_1) &= 0 \qquad \text{if } n > 1. \end{aligned}$$

(2) We say that a continuous function $f : X \to Y$ is a *Serre fibration* if it satisfies the homotopy lifting property: this means that for any compact polyhedron P and any continuous function $h : P \times I \to Y$ such that $h_0 : x \mapsto h(x, 0)$ lifts to a continuous function $k_0 : P \to X$, there exists a continuous function $k : P \times I \to X$ such that $f \circ k = h$ and $k(x, 0) = k_0$ for all $x \in P$:

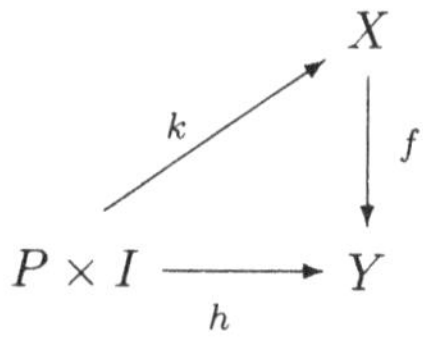

Let $f : X \to Y$ be a Serre fibration, $a \in X$, $b = f(a)$ and $F = f^{-1}(b)$ the fibre of f over the point b. Then we can construct a long exact sequence

$$\cdots \to \pi_n(F, a) \to \pi_n(X, a) \to \pi_n(Y, b) \xrightarrow{\delta} \pi_{n-1}(F, a) \to \cdots$$

$$\to \pi_0(F, a) \to \pi_0(X, a) \to \pi_0(Y, b),$$

where $\pi_0(X, a)$ is the set of connected components of X considered as a topological space with base point.

Examples.

(1) Let $f : X \to Y$ be a locally trivial topological fibration with fibre F. Then f is a Serre fibration. In particular, if X and Y are C^∞ manifolds and $f : X \to Y$ a submersion which is proper and surjective then locally we have $X|_U \cong U \times F$, where F is the fibre of f over the point b: we say that f is *locally trivial.* It follows that such a map is a Serre fibration.

(2) Consider the Hopf fibration

$$S_3 \longrightarrow S_2 = \mathbf{P}_1(\mathbb{C})$$

induced by the projection $\mathbb{C}^2 \setminus \{0\} \to \mathbf{P}_1(\mathbb{C})$. The preceding exact sequence shows that $\pi_3(S_3) = \pi_3(S_2) = \mathbb{Z}$.

(3) The Poincaré group $\pi_1(X, a)$ acts on $\pi_n(X, a)$ in the following way. Let $\alpha : (I, \partial I) \to (X, a)$ and $\beta : (I^n, \partial I^n) \to (X, a)$ be two continuous functions. Then there exists a continuous function $h : I^{n+1} \to X$ such that $h(x, t) = \alpha(t)$ for $x \in \partial I^n$ and $t \in I$, and $h(x, 0) = \beta(x)$. The function $x \mapsto h(x, 1)$ defines a homotopy class $(I^n, \partial I^n) \to (X, a)$ which only depends on the homotopy classes $[\alpha]$ and $[\beta]$, and which we shall denote by $[\alpha].[\beta]$. This action is compatible with the group structure of $\pi_n(X, a)$, in other words, we have

$$[\alpha].([\beta_1].[\beta_2]) = ([\alpha].[\beta_1])([\alpha].[\beta_2])$$

for $[\alpha] \in \pi_1(X, a)$, and $[\beta_1]$ and $[\beta_2] \in \pi_n(X, a)$.

Proposition 3.1.1 *If the topological space X is arc-connected then the set of orbits of $\pi_1(X, a)$ in $\pi_n(X, a)$ can be identified with the set $[S_n, X]$ of homotopy classes of continuous maps from the sphere S_n to X.*

Proposition 3.1.2 *Let G be a topological group with identity element e. The action of $\pi_1(G, e)$ on $\pi_n(G, e)$ defined above is trivial.*

If G is connected then the group $\pi_n(G, e)$ coincides with $[S_n, G]$; the group law being induced by that of G.

Example.

The inclusion $U(r) \hookrightarrow \mathrm{GL}(r, \mathbb{C})$ is a homotopy equivalence. The homotopy groups $\pi_i(U(r))$ can be calculated for $i \leq 3$ using the Serre fibrations

$$\begin{aligned} U(1) &\longrightarrow U(2) \longrightarrow S_3, \\ U(2) &\longrightarrow U(3) \longrightarrow S_5, \\ &\vdots \\ U(r-1) &\longrightarrow U(r) \longrightarrow S_{2r-1}. \end{aligned}$$

We obtain

$$\begin{aligned} \pi_1(U(r)) &\simeq \mathbb{Z}, \\ \pi_2(U(r)) &= 0, \\ \pi_3(U(r)) &\simeq \mathbb{Z}. \end{aligned}$$

For $i > 3$ the situation is more delicate because the higher index homotopy groups of spheres play a role. For example, $\pi_i(U(2)) = \pi_i(S_3)$ if $i \geq 4$. Note that, for $i < 2r$, we have $\pi_i(U(r)) \simeq \pi_i(U(r+1))$. The groups

$$\pi_i(U) = \varinjlim_r \pi_i(U(r))$$

are called *the stable homotopy groups* of the unitary group. It is the Bott Periodicity Theorem which enables us to compute these groups.

3.2 Vector bundles on a contractible space

Let X be a topological space which, for simplicity, we assume to be compact. Let $\mathrm{Vect}_r(X)$ denote the set of isomorphism classes of topological $\mathbb{C}$-vector bundles of rank r on X.

Proposition 3.2.1 *Let E be a topological vector bundle on a contractible compact topological space X. Then E is trivial.*

Lemma 3.2.2 *Let E be a vector bundle on $X \times I$. For each $t \in I$ let E_t denote the vector bundle induced on X by the inclusion $X \hookrightarrow X \times I$ defined by $x \mapsto (x, y)$. Then*

$$E_0 \simeq E_1.$$

Proof Let $p : X \times I \to X$ be the projection to the first factor. Over $X \times \{t\}$ the bundles $p^*(E_t)$ and E coincide. It follows that we can find an isomorphism $p^*(E_t) \simeq E$ over a neighbourhood $X \times V$ of $X \times \{t\}$ and for every $t' \in V$ we have $E_{t'} \simeq E_t$. Hence, the isomorphism class of E_t is locally constant on I and so constant. □

Lemma 3.2.3 *Let f and $g : X \to Y$ be two homotopic maps and E a topological vector bundle on Y. Then*

$$f^*(E) \simeq g^*(E).$$

Proof Let $h : X \times I \to Y$ be a homotopy between f and g. Then we have $h(x,0) = f(x)$ and $h(x,1) = g(x)$ for all $x \in X$. It is enough to apply Lemma 3.2.2 to the bundle $h^*(E)$ over $X \times I$. □

Proof of Proposition 3.2.1 To say that the topological space X is contractible means that there exists a point $a \in X$ such that the map id_X is homotopic to the constant map $i_a : X \to X$ given by the point a. Hence,

$$E \simeq i_a^*(E) = X \times E_a.$$

□

3.3 Gluing vector bundles

Let $X = X_1 \cup X_2$ be a cover of X by two closed sets X_1 and X_2. We put $X_{12} = X_1 \cap X_2$. Let $\varphi : X_{12} \to \mathrm{GL}(r, \mathbb{C})$ be a continuous function. We define a vector bundle $E(\varphi)$ on X by setting

$$E(\varphi) = ((X_1 \times \mathbb{C}^r) \coprod (X_2 \times \mathbb{C}^r))/R,$$

where R is the finest equivalence relation which identifies $(x_1, v_1) \in X_1 \times \mathbb{C}^r$ and $(x_2, v_2) \in X_2 \times \mathbb{C}^r$ if $x_1 = x_2$ and $v_2 = \varphi(x_1)v_1$. Let $\varphi' : X_{12} \to \mathrm{GL}(r, \mathbb{C})$ be another continuous function homotopic to φ. Using the homotopy

$$h : X_{12} \times I \longrightarrow \mathrm{GL}(r, \mathbb{C})$$

between φ and φ' we can construct, in the same way as before, a vector bundle on $X \times I$ which coincides with $E(\varphi)$ on $X \times \{0\}$ and with $E(\varphi')$ on $X \times \{1\}$. Then by Lemma 3.2.2 we have $E(\varphi) \simeq E(\varphi')$. It follows that we obtain a map

$$[X_{12}, \mathrm{GL}(r, \mathbb{C})] \longrightarrow \mathrm{Vect}_r(X)$$

which associates to a homotopy class $[\varphi]$ the isomorphism class of the vector bundle $E(\varphi)$.

Proposition 3.3.1 *If X_1 and X_2 are contractible then this map is bijective.*

Proof 1. *Surjectivity.* Let E be a vector bundle on X. By Proposition 3.2.1 there exist trivializations

$$\varphi_1 : E|_{X_1} \simeq X_1 \times \mathbb{C}^r \text{ and } \varphi_2 : E|_{X_2} \simeq X_2 \times \mathbb{C}^r.$$

Then on X_{12} we get an isomorphism of trivial vector bundles $\varphi_2\varphi_1^{-1} : X_{12} \times \mathbb{C}^r \to X_{12} \times \mathbb{C}^r$. This map is given by a continuous function $\varphi : X_{12} \to \mathrm{GL}(r, \mathbb{C})$ and we clearly have $E(\varphi) \simeq E$.

2. *Injectivity.* Let φ and $\varphi' : X_{12} \to \mathrm{GL}(r, \mathbb{C})$ be two continuous functions such that $E(\varphi) \simeq E(\varphi')$. Then there exist two continuous maps

$$g_1 : X_1 \longrightarrow \mathrm{GL}(r, \mathbb{C}) \text{ and } g_2 : X_2 \longrightarrow \mathrm{GL}(r, \mathbb{C})$$

such that we have $\varphi' = g_2\varphi g_1^{-1}$ on X_{12}. If X_1 and X_2 are contractible then the maps g_1 and g_2 are homotopic to the constant map given by the identity element $\mathrm{id}_{\mathbb{C}^r}$. It follows that φ is homotopic to φ', from which we deduce the proposition. □

Corollary 3.3.2 *We have*

$$\mathrm{Vect}_r(S_n) \simeq \pi_{n-1}(\mathrm{GL}(r, \mathbb{C})).$$

Corollary 3.3.3 *Let X be a compact topological space and SX the suspension of X. Then*

$$\mathrm{Vect}_r(SX) \simeq [X, \mathrm{GL}(r, \mathbb{C})].$$

Exercise

Let X be a locally compact topological space which is countable at infinity. Show that if X is contractible then every vector bundle on X is trivial. (Generalize Lemma 3.2.2 to this situation by writing the topological space X as a countable union of compact spaces.)

3.4 The Riemann surface case

We consider the compact subset P of $\mathbb{C} = \mathbb{R}^2$ bounded by the regular polygon with $4g$ vertices $s_k = \exp(ik\pi/2g)$. Let a_k, b_k, a_k^{-1} and b_k^{-1} denote the sides $s_{k-1}s_k$ given by $k \equiv 1, 2, 3$ or $4 \bmod 4$, oriented in the positive sense or otherwise according to whether the exponent is 1 or -1. We consider the quotient topological space X of P by the equivalence relation which identifies the oriented side a_k with the oriented side a_k^{-1} and the oriented side b_k with the oriented side b_k^{-1}. Let $p : P \longrightarrow X$ be the canonical projection. The topological space X is a compact connected real surface and its Poincaré group can be obtained by applying the Van Kampen Theorem. Indeed, one can cover P by the open sets

$$P' = \{z \in P||z| > \frac{1}{3}\} \text{ and } P'' = \{z \in P||z| < \frac{2}{3}\}$$

and we denote their images in X by X' and X''. The open set X' has the homotopy type of a bouquet of $2g$ circles and the open set X'' is contractible. Moreover, the intersection $Y = X' \cap X''$ has the homotopy type of a circle. Let $a = p(\frac{1}{2})$. The Van Kampen Theorem shows that the Poincaré group is the amalgamated sum of the Poincaré groups of X' and X'' given by the diagram

$$\begin{array}{ccc} \pi_1(Y, a) & \longrightarrow & \pi_1(X'', a) \\ \downarrow & & \\ \pi_1(X', a) & & \end{array}$$

Let α_k be the class of the loop $p(a_k)$ and β_k the class of the loop $p(b_k)$. Then the Poincaré group of X can be identified with the quotient of the free group $\mathcal{L}(\alpha_1, \beta_1, \ldots, \alpha_{2g}, \beta_{2g})$ by the distinguished subgroup generated by $c = \prod_{k=1,\ldots,2g}[\alpha_k, \beta_k]$, where we have set $[\alpha_k, \beta_k] = \alpha_k\beta_k\alpha_k^{-1}\beta_k^{-1}$.

We can use this cover and the Mayer-Vietoris exact sequence [27] to determine the Betti numbers of X. We obtain the exact sequence

$$\cdots \longrightarrow H^1(X) \longrightarrow H^1(X') \xrightarrow{\rho} H^1(Y) \longrightarrow H^2(X) \longrightarrow 0.$$

It is clear that the map ρ vanishes and it follows that the Betti numbers of X are $b_1 = 2g$ and $b_2 = 1$. Hence, the surface X is an orientable surface of genus g. It is well known (cf. [25]) that all compact connected surfaces are homeomorphic to such surfaces.

Vector bundles

We let $\mathrm{Pic}(X)$ denote the group of isomorphism classes of topological vector bundles of rank one on X and we let $\mathbb{C}^r_X$ denote the trivial bundle of rank r.

Theorem 3.4.1 *(a) The group* $\mathrm{Pic}(X)$ *is isomorphic to* $\mathbb{Z}$.
(b) The map

$$L \mapsto L \oplus \mathbb{C}_X^{r-1} : \mathrm{Pic}(X) \longrightarrow \mathrm{Vect}_r(X)$$

is bijective.

Proof Consider the cover of P by the closed sets $A' = \{z \in P \| z| \geq \frac{1}{2}\}$ and $B' = \{z \in P \| z| \leq \frac{1}{2}\}$, and let A and B denote their images in X. The closed set B is still contractible and A has the homotopy type of a bouquet of $2g$ circles. Let C be the circle $A \cap B$. Every complex vector bundle is trivial on A and on B so that we obtain, as in Proposition 3.3.1, a surjective map

$$[C, \mathrm{GL}(r, \mathbb{C})] \to \mathrm{Vect}_r(X)$$

which identifies $\mathrm{Vect}_r(X)$ with the quotient of the group $[C, \mathrm{GL}(r, \mathbb{C})]$ by the right action of the group $[A, \mathrm{GL}(r, \mathbb{C})]$ via the function homomorphism from A to C. For $r = 1$, this homomorphism is trivial and we get that $\mathrm{Pic}(X)$ is in bijective correspondence with the group $[C, \mathbb{C}^*]$ which is isomorphic to $\mathbb{Z}$. The group law is induced by multiplication in $\mathbb{C}^*$ and it is clear that this bijection is actually a group isomorphism. For the other part, consider the inclusion $\mathbb{C}^* \hookrightarrow \mathrm{GL}(r, \mathbb{C})$ defined by

$$\lambda \longmapsto \begin{pmatrix} \lambda & 0 & \cdots & 0 \\ 0 & 1 & \cdots & 0 \\ & & \ddots & \\ & \cdots & 0 & 1 \end{pmatrix}.$$

We have seen in section 3.1 that this inclusion induces an isomorphism of Poincaré groups. The commuting diagram

$$\begin{array}{ccc} [C, \mathbb{C}^*] & \longrightarrow & \mathrm{Pic}(X) \\ \downarrow & & \downarrow \\ [C, \mathrm{GL}(r, \mathbb{C})] & & \mathrm{Vect}_r(X) \end{array}$$

shows that the map $\mathrm{Pic}(X) \to \mathrm{Vect}_r(X)$ defined by $L \mapsto F = L \oplus \mathbb{C}_X^{r-1}$ is surjective. It is trivially injective because $\wedge^r F \simeq L$. □

3.5 The degree, as topological invariant

We keep to the same notation as used in the previous section.

Lemma 3.5.1 *Let L be a topological vector bundle of rank 1 over X. Then L has a section which has only isolated zeros.*

Proof We can suppose that the bundle L is given by the homotopy class of the function $\gamma : C \to \mathbb{C}^*$ defined by $\gamma(z) = z^k$ for $z \in \mathbb{C}$. Then L has non-zero sections s and t over A and B respectively and, over C, we have the relation $s(z) = \gamma(z)t(z)$. The function γ extends to a continuous function $\overline{\gamma}$ on B which only vanishes at the origin. Then, there exists a continuous section σ of L defined by

$$\sigma = \begin{cases} s & \text{on } A, \\ \overline{\gamma}t & \text{on } B. \end{cases}$$

The zeros of this section coincide with those of the function $\overline{\gamma}$ and so it has only one zero. □

Consider a section s of a topological vector bundle L over X and suppose that s has only isolated zeros. Let a be one of these zeros and choose a neighbourhood D of a homeomorphic to a ball on which the bundle has a non-zero section e and on which s has no zeros other than a. We write $s(z) = f(z)e(z)$; since $D \setminus \{a\}$ has the homotopy type of a circle, the function $f : D \setminus \{a\} \to \mathbb{C}^*$ has a degree which is independent of the choice of D and of the section e: this number is called the *local degree* of s at a denoted by $\deg_a(s)$. We let $Z(s)$ denote the set of zeros of s and put

$$\deg(s) = \sum_{a \in Z(s)} \deg_a(s).$$

Theorem 3.5.2 *The degree of a continuous section s of L with only isolated zeros is independent of s.*

We let $d(L)$ denote this number: it is the *degree* of L. The section constructed in Lemma 3.5.1 can be used to show that the degree of a bundle of rank 1 associated to a map $C \to \mathbb{C}^*$ of degree k is k. Thus, the isomorphism $\mathrm{Pic}(X) \simeq \mathbb{Z}$ of Theorem 3.4.1 is canonical.

Corollary 3.5.3 *The map $\mathrm{Pic}(X) \to \mathbb{Z}$ given by $L \mapsto d(L)$ is a group isomorphism.*

Any topological vector bundle F of rank r has an associated bundle of rank one given by $\det(F) = \Lambda^r(F)$. We set $d(F) = d(\det(F))$. Taking Theorem 3.4.1 into account, we obtain

Corollary 3.5.4 *The map* $\mathrm{Vect}_r(X) \to \mathbb{Z}$ *defined by* $F \mapsto d(F)$ *is bijective.*

The irreducible smooth algebraic projective curves can be viewed as compact connected orientable real surfaces. To each algebraic vector bundle on such a curve we can associate a topological bundle.

Corollary 3.5.5 *Let F be an algebraic vector bundle of rank r over a connected smooth projective curve X. Then the degree of F, defined in section 2.6, coincides with the degree of the underlying topological bundle.*

Proof By virtue of Theorem 2.6.9, we can reduce to the case of a rank 1 bundle L. Then we can suppose that L has a regular section s which is not identically zero. The zeros of s are isolated and the local degree calculated at a zero is the multiplicity of that zero. Thus, the degree s is the degree of the zero divisor of s. □

Proof of Theorem 3.5.2 We may assume that the surface X has the structure of a C^∞ manifold of dimension 2. If a is a point of X then the open set $X \setminus \{a\}$ has the homotopy type of a bouquet of $2g$ circles and the argument that we used in the proof of Theorem 3.4.1 shows that any topological bundle of rank one is isomorphic to a C^∞ bundle of rank one. We suppose that the bundle given by L is C^∞.

Lemma 3.5.6 *The C^∞ sections of L are dense in the space $C^0(X, L)$ of continuous sections.*

To give meaning to this statement we must say precisely which topology we are considering on the space of continuous sections. We can consider a Hermitian metric on each fibre L_x of L which we ask to depend continuously on x: this makes sense if we choose local trivializations of the bundle L. Then we put

$$||s|| = \sup_{x \in X} ||s(x)||.$$

Now, we can cover X by a finite number of open sets $(V_i)_{i=1,\dots,k}$ which are relatively compact in the open sets U_i over which L is trivial. Consider a partition of unity φ_i subordinate to this cover (V_i). Let s be a continuous section of L and ϵ a positive real number. Then we can find, using the Weierstrass Density Theorem, a section s_i of class C^∞ over V_i such that

$$||s - s_i||_{V_i} \leq \epsilon.$$

Consider the C^∞ section of L defined by $u = \sum_i \varphi_i s_i$. Then we have $||s - u|| \leq \epsilon$.

Lemma 3.5.7 *The C^∞ sections which are transverse to the zero section of L are dense in the vector space of C^∞ sections.*

Proof The differential $d_x s : T_x X \to L_x$ of a C^∞ section at a point $x \in X$ where it vanishes is defined in the following way. Choose a non-zero section e over a neighbourhood V of x, write $s = fe$ over the open set V, where f is a C^∞ function on V, and put $d_x s = (d_x f)e$. One can check that this linear map does not depend on the trivialization chosen in a neighbourhood of x. To say that such a C^∞ section is transverse to the zero section means that at every point $x \in X$ where it vanishes, the differential

$$d_x s : T_x X \longrightarrow L_x$$

surjects.

Consider a C^∞ section u and a vector space Γ of finite dimension k of C^∞ sections containing u which generates the bundle L. The evaluation map gives rise to a surjective map

$$X \times \Gamma \longrightarrow L$$

of complex vector bundles. Denote the kernel by N. The total space of N is then a smooth real manifold of dimension k. Consider the projection $q : N \to \Gamma$. It is clear that the sections transverse to the zero section correspond to points $s \in \Gamma$ which are not critical values of the map q. Sard's Theorem implies that the set of critical values of q has no interior points. The statement follows from this. □

Note that if a C^∞ section s is transverse to the zero section then its zeros are isolated. So we can say what the degree of such a section is. If b is one of the zeros of s then the local degree of s at b is calculated in the following way. Let $\overline{B}$ be a small compact ball centred on b in which the section has b as its only zero. In a neighbourhood of this ball we trivialize L using a non-zero section e, which is possible if the ball is small enough; we write $s = ef$ is a neighbourhood of $\overline{B}$. Then

$$\deg_b s = \frac{1}{2i\pi}\int_{\partial B}\frac{df}{f}.$$

Lemma 3.5.8 *Let u be a continuous section of L which has only isolated zeros. Then there exists a section v of class C^∞ of L which is transverse to the zero section and such that* $\deg u = \deg v$.

Proof Let $\{a_1, \dots, a_\ell\}$ be the set of zeros of u. For each integer j ($j = 1, \dots, \ell$) choose an open ball B_j centred on a_j; we can assume that these balls are disjoint and that over each neighbourhood of $\overline{B}_j$ the bundle L has a C^∞ section e_j which has no zeros. Let $K = X \setminus \bigcup_{j=1,\dots,\ell} B_j$. Consider the closed subset of $C^0(X, L) \times X$ given by the pairs (v, x) such that $v(x) = 0$, $x \in K$. By definition, the image of this closed set in $C^0(X, L)$ is a closed set which does not contain u. Hence, every section v of L sufficiently near to u has its zeros in the balls B_j; write $v|_{B_j} = v_j e_j$, where $v_j : B_j \to \mathbb{C}^*$ is a continuous function. By homotopy, we have

$$\deg_{a_j} u = \deg(v_j|_{\partial B_j}).$$

By Lemmas 3.5.6 and 3.5.7 we can choose such a smooth section v which is transverse to the zero section. Then v_j is C^∞ and we have

$$\deg(v_j|_{\partial B_j}) = \frac{1}{2i\pi}\int_{\partial B_j}\frac{dv_j}{v_j}.$$

Let $\{b_{j,1}, \dots, b_{j,\nu_j}\}$ be the zeros of v in B_j. If $B_{j,k}$ ($k = 1, \dots, \nu_j$) are small disjoint balls contained in B_j and centred on $b_{j,k}$ then we have

$$\deg(v_j|_{\partial B_j}) = \frac{1}{2i\pi}\sum_{k=1,\dots,\nu_j}\int_{\partial B_{j,k}}\frac{dv_j}{v_j}$$

by Stokes' formula. It follows that $\deg u = \deg v$. □

It remains to prove that the degree of a C^∞ section which is transverse to the zero section does not depend on the section. We use the notion of Chern classes.

3.6 Connections and Chern classes

Connections

Consider a rank one C^∞ vector bundle L on a C^∞ manifold X. Let $A^i(L)$ denote the vector space of differential forms of class C^∞ with values in L. By a *connection* on L we mean a $\mathbb{C}$-linear map $\nabla : A^0(L) \to A^1(L)$ which satisfies the Leibniz condition $\nabla(fs) = df.s + f\nabla s$, where $f : X \to \mathbb{C}$ is a C^∞ function and $s \in A^0(L)$.

Lemma 3.6.1 *(i) There exist connections on L.*

(ii) In the space of $\mathbb{C}$-linear operators $A^0(L) \to A^1(L)$ the connections form an affine sub-variety with tangent space given by the vector space $A^1(X)$ of differential forms on X of degree 1.

Proof Consider a locally finite open cover (U_i) on which the bundle L is trivial. We consider a partition of unity (φ_i) subordinate to the cover (U_i) and, for each integer, a non-zero section e_i of L over the open set U_i. If s is a C^∞ section of L then we write $s|_{U_i} = s_i e_i$ and put

$$\nabla s = \sum_i \varphi_i ds_i \, e_i.$$

This defines a connection on L. Now, the difference D of two connections is a linear operator over the algebra $C^\infty(X)$ of C^∞ functions on the manifold X: in other words, $D(fs) = fD(s)$ for $f \in C^\infty(X)$ and $s \in A^0(L)$. Such an operator gives a map of bundles $L \to T^* \otimes L$; since the bundle L has rank one and this map is given by multiplication by a scalar differential form. □

Chern classes

Consider a connection ∇ on the bundle L. This connection extends to an operator which we still denote by $\nabla : A^i(L) \to A^{i+1}(L)$: it is defined by requiring that it satisfies the Leibniz formula

$$\nabla(\alpha s) = d\alpha.s + (-1)^i \alpha \wedge \nabla s$$

for $\alpha \in A^i(X)$ and $s \in A^0(L)$. The composite operator $\nabla^2 : A^i(L) \to A^{i+2}(L)$ is linear over the algebra $C^\infty(X)$ and so is exterior multiplication by a scalar differential form of degree 2 which we write as

$$\nabla^2(s) = \omega s$$

for $s \in A^0(L)$. The differential form ω is called the *curvature form* of the connection ∇.

Lemma 3.6.2 *The differential form ω is closed. In de Rham cohomology its class is independent of the connection.*

Proof The first part is obtained by computing ∇^2. For the second part, note that the connection $\nabla + \alpha$, with $\alpha \in A^1(X)$, has curvature form $\omega + d\alpha$. □

The class in $H^2_{\mathrm{DR}}(X)$ of the form $\frac{i}{2\pi}\omega$ is called the *Chern class* of L and is denoted $c(L)$. One can check easily that the map $c : \mathrm{Pic}(X) \to H^2_{\mathrm{DR}}(X)$ given by $L \mapsto c(L)$ is a homomorphism of groups. Theorem 3.5.2 is a consequence of the following statement which also justifies the introduction of the coefficient $\frac{i}{2\pi}$.

Proposition 3.6.3 *Let X be a compact Riemann surface, L a C^∞ vector bundle of rank 1 on X and s a C^∞ section of L transverse to the zero section. Then*

$$\deg s = \int_X c(L).$$

Proof Let $Z = \{a_1, \ldots, a_\ell\}$ be the set of zeros of s. Over the open set y$U = X \setminus Z$ we write $\nabla s = \alpha s$, where α is a C^∞ form of degree 1 over U. Then, the curvature form is $\omega = d\alpha$ over the open set U. Let $\overline{B}_j$ be a small closed ball centred on a_j of radius ρ; we suppose that the balls are disjoint and that the bundle L has a section e_j which does not vanish in a neighbourhood of $\overline{B}_j$. By Stokes' formula we have

$$\begin{aligned}\int_X c(L) &= \frac{i}{2\pi}\int_X \omega \\ &= \lim_{\rho\to 0}\frac{i}{2\pi}\int_{X\setminus\bigcup B_j} d\alpha \\ &= \sum_{j=1}^{\ell}\lim_{\rho\to 0}\Big(\frac{1}{2i\pi}\int_{\partial B_j}\alpha\Big).\end{aligned}$$

We write $s = fe_j$ in a neighbourhood of $\overline{B}_j$, where f is a scalar function which only vanishes at the point a_j. Then we have

$$\alpha = \frac{df}{f} + \beta,$$

where β is a C^∞ form in a neighbourhood of $\overline{B}_j$. It follows that

$$\lim_{\rho\to 0}\frac{1}{2i\pi}\int_{\partial B_j}\alpha = \deg_{a_j} s.$$

Therefore, $\int_X c(L) = \deg s$, which completes the proof. □

4

The Hilbert Scheme

The most important result of this chapter is the construction of the Hilbert scheme – or the Hilbert-Grothendieck scheme – which parametrizes the coherent sheaves with a given Hilbert polynomial which are quotients of a given locally-free sheaf E. The method we use is that of Grothendieck [29] and amounts to identifying this set with a locally closed sub-variety of a Grassmannian. However, while Grothendieck's construction is useful in all dimensions, we will only tackle the case where the base variety is a curve X; this considerably reduces the technical difficulties. The Hilbert-Grothendieck scheme is characterized by a universal property which makes the introduction of flat families $(F(s))_{s \in S}$ of coherent sheaves on X parametrized by an algebraic variety S indispensable. When the parameter variety S is reduced this notion can be characterized in terms of global invariants and simply means that the Hilbert polynomial of $F(s)$ (that is, the rank and degree) remains constant. When the variety S is not reduced we can characterize the flatness by introducing direct image sheaves. The first three sections of this chapter are devoted to studying these two ideas; for more details, consult sections 8 and 9 of chapter III in Hartshorne's book [31].

4.1 Flatness

Let A be a (commutative, unitary) ring and M an A-module. We say that M is *flat* over A if the functor, defined on the category of A-modules, $N \mapsto M \otimes_A N$ is exact. This is equivalent to the condition

$$\mathrm{Tor}_1^A(M, N) = 0$$

for all A-modules N. A projective module is obviously flat. In the case of a module of finite type over a local ring we have the following flatness criterion:

Proposition 4.1.1 *Let A be a local ring with maximal ideal $\mathfrak{m}$ and residue field $k = A/\mathfrak{m}$. An A-module M of finite type is flat if and only if*

$$\mathrm{Tor}_1^A(M, k) = 0.$$

This proposition is another version of Lemma 2.3.3. The following is another criterion which is also very useful but its proof is harder [9, Chapter III, §5, Theorem 1]:

Proposition 4.1.2 *Let $\varphi : A \to B$ be a local map of local Noetherian rings with maximal ideals $\mathfrak{m}$ and $\mathfrak{n}$, respectively, and let M be a B-module of finite type. Let $k = A/\mathfrak{m}$ be the residue field. Then M is A-flat if and only if*

$$\mathrm{Tor}_1^A(M, k) = 0.$$

Corollary 4.1.3 *Under the same hypotheses, if B is flat over A then M is a flat A-module if and only if*

$$\mathrm{Tor}_1^B(M, B/\mathfrak{m}B) = 0.$$

Consider a morphism of algebraic varieties $f : Y \to S$. A coherent algebraic sheaf F on Y is said to be S-*flat* if, for each point $y \in Y$, the module F_y is flat over the local ring $A = \mathcal{O}_{S,s}$, where $s = f(y)$, and where F_y is considered as an A-module via the map of local rings $f^* : A = \mathcal{O}_{S,s} \to \mathcal{O}_{Y,y} = B$. If the sheaf $F = \mathcal{O}_Y$ is S-flat then we say that Y is S-flat. If Y is S-flat and if we have a locally-free left resolution

$$\mathcal{L}_\bullet \longrightarrow F$$

of F then the flatness condition is equivalent, by the last proposition, to the following condition: for every point $s \in S$, the induced map

$$\mathcal{L}_\bullet(s) \longrightarrow F(s)$$

on the fibre $Y(s)$ over the point s is a resolution of $F(s)$.

4.2 Direct images

Let $f : Y \to S$ be a morphism of algebraic varieties and F a coherent sheaf on Y. The presheaf which associates to every open subset U of S the module $\Gamma(f^{-1}(U), F)$ is an algebraic sheaf on S called the *direct image* of F and we denote it by $f_*(F)$. The functor $F \mapsto f_*(F)$ is left exact; its right derived functors are denoted $R^q f_*(F)$ and are called the *higher direct images* of F.

Lemma 4.2.1 *The sheaf $R^q f_*(F)$ can be identified with the associated sheaf of the presheaf which takes the open subset U of S to the module $H^q(f^{-1}(U), F)$.*

Proof Pick an injective resolution $F \to R^\bullet$ of F. By definition, the direct image $R^q f_*(F)$ is the qth cohomology sheaf of the complex $f_*(R^\bullet)$; hence, it must be the sheaf associated to the presheaf given by $U \mapsto H^q(\Gamma(U, f_*(R^\bullet))) = H^q(\Gamma(f^{-1}(U), R^\bullet))$. The statement follows from this. □

Proposition 4.2.2 *Let X be an algebraic variety, p and q the canonical projections $S \times X \to S$ and $S \times X \to X$. Let G be a coherent algebraic sheaf on X and $F = q^*(G)$. Then, for each integer j, the canonical map*

$$\mathcal{O}_S \otimes H^j(X, G) \longrightarrow R^j p_*(F)$$

is an isomorphism.

Proof We construct such a map by considering an injective resolution $G \to R^\bullet$ of the sheaf G. By pullback, we obtain a resolution $F \to q^*(R^\bullet)$ and hence a map

$$H^j(\mathcal{O}_S \otimes \Gamma(X, R^\bullet)) \longrightarrow R^j p_*(F).$$

The domain can be identified with $\mathcal{O}_S \otimes H^j(X, G)$. It remains to see that this map is an isomorphism. The question is local on S; so we can suppose that S is an affine variety. If X is also an affine variety then the result is obvious. In the case where the variety X is arbitrary, we can use the following lemma which is the analogy in sheaf theory of the Mayer-Vietoris exact sequence.

Lemma 4.2.3 *Let X_1 and X_2 be two open subsets of a topological space X which cover X and have intersection X_{12}. For any sheaf F of abelian* *groups on X we have a long exact sequence*

$$\cdots \to H^j(X,F) \to H^j(X_1,F) \oplus H^j(X_2,F)$$
$$\to H^j(X_{12},F) \to H^{j+1}(X,F) \to \cdots$$

It follows from the Five Lemma that if the proposition is true for X_1, for X_2 and for X_{12} then it is true for X. Consequently, it suffices to cover the variety X by a finite union of affine open subsets U_i. The proposition is true when we replace X by any of the open subsets and by the intersections of these open subsets; then, by induction on the number of open subsets, we see that it is true for all finite unions of such open subsets. □

Proof of the lemma. Consider an injective resolution $F \to R^\bullet$ of F. Such a resolution is flasque so we have an exact sequence of complexes

$$0 \longrightarrow \Gamma(X,R^\bullet) \longrightarrow \Gamma(X_1,R^\bullet) \oplus \Gamma(X_2,R^\bullet) \longrightarrow \Gamma(X_{12},R^\bullet) \longrightarrow 0$$

and it suffices to take the associated long exact sequence. □

The fundamental theorems

Suppose that the morphism $f : Y \to S$ is a projective morphism. Let $\mathcal{O}_Y(1)$ denote the invertible sheaf induced by the embedding $Y \hookrightarrow S \times \mathbf{P}_k$ into the relative projective space over S. For each algebraic sheaf F on Y we put $F(n) = F \otimes \mathcal{O}_Y(n)$.

Lemma 4.2.4 *Let $f : Y \to S$ be a projective morphism over an affine variety S and let F be a coherent algebraic sheaf on Y. Then the sheaf F is a quotient of a sheaf of the type $(\mathcal{O}_Y(-n))^k$.*

Proof We may assume that $Y = S \times X$, where X is a projective space and f is the projection to the first factor. If U_i is a finite family of affine open subsets of X which cover X then we can find sections $s_{i,j}$ which generate F over the affine open set $S \times U_i$. We may assume that the complement of U_i is defined by an equation $u_i = 0$, where u_i is a section of $\mathcal{O}_X(m)$ for a suitable m. Then, for ℓ large enough, the sections $s_{i,j}u_i^\ell$ of $F(m\ell)$ extend over $S \times X$ and the canonical map

$$\mathcal{O}_Y^k \longrightarrow F(m\ell)$$

which they define surjects. □

This lemma has many important consequences.

Theorem 4.2.5 *Let $f : Y \to S$ be a projective morphism and F a coherent algebraic sheaf on Y. Then the algebraic sheaves $R^j f_*(F)$ are coherent.*

Proof The question is local on S so we can assume that it is affine. Furthermore, we can assume that $Y = S \times X$, where X is a projective variety. We argue by reverse induction on j. If $j \gg 0$ then the direct image vanishes. Suppose that the statement is true in degree $j+1$. The sheaf F fits into an exact sequence $0 \to G \to \mathcal{L} \to F \to 0$, where the sheaf $\mathcal{L}$ is isomorphic to $(\mathcal{O}_Y(-n))^k$. It follows from Proposition 4.2.2 that the direct images $R^j f_*(\mathcal{L})$ are trivial sheaves. From the exact sequence

$$R^j f_*(G) \to R^j f_*(\mathcal{L}) \to R^j f_*(F) \to R^{j+1} f_*(G) \to R^{j+1} f_*(\mathcal{L})$$

and the induction hypothesis, we deduce that $R^j f_*(F)$ is locally of finite type. This clearly applies to the sheaf G; it follows from this that the cokernel of the first map is coherent and so the sheaf $R^j f_*(F)$ is itself coherent. □

Lemma 4.2.4 can be used to prove relative versions of Serre's Theorems A and B.

Theorem 4.2.6 *Relative Theorem A Let $f : Y \to S$ be a projective morphism. If F is a coherent algebraic sheaf on Y then there exists an integer ν such that, for $m \geq \nu$, the canonical evaluation map*

$$f^*(f_*(F(m))) \longrightarrow F(m)$$

surjects.

Proof The question is local on S. We can assume that S is affine, that $Y = S \times X$, where X is a projective variety, and that f is the projection to the first factor. The theorem hold trivially for $\mathcal{O}_Y(k)$ by Proposition 4.2.2 and Theorem A for projective space. To obtain the theorem for an arbitrary coherent sheaf F, we can suppose that F is a quotient of a sheaf L which is a direct sum of invertible sheaves $\mathcal{O}_Y(k)$.

Then the commuting diagram

$$\begin{array}{ccc} f^*(f_*(L(m))) & \longrightarrow & L(m) \\ \downarrow & & \downarrow \\ f^*(f_*(F(m))) & \longrightarrow & F(m) \end{array}$$

immediately leads us to the result □

Theorem 4.2.7 *Relative Theorem B Let $f : Y \to S$ be a projective morphism and let F be a coherent algebraic sheaf on Y. There exists an integer ν such that, for $m \geq \nu$ and $q > 0$, we have $R^q f_*(F(m)) = 0$.*

Proof We can still assume that S is affine and , since the map f is projective, we can assume that $Y = S \times X$, where X is a projective variety and f is the projection to the first factor. The statement is trivial for the sheaves $\mathcal{O}_Y(k)$ for $k \in \mathbb{Z}$ by Proposition 4.2.2. In the case where F is an arbitrary coherent sheaf we use reverse induction on q and the exact sequence $0 \to G \to L \to F \to 0$ of coherent algebraic sheaves, where L is a direct sum of appropriate invertible sheaves $\mathcal{O}_Y(k)$. □

Description of direct images

From now on we assume that $Y = S \times X$, where X is a smooth projective curve and we take the projection $p : S \times X \to S$ as f. Let F be a coherent algebraic sheaf on Y and let $F(s)$ denote the algebraic sheaf on X induced by the inclusion $X \hookrightarrow S \times X$ given by the point s.

If S is affine then, by Lemma 4.2.4, the sheaf F has a left resolution $\mathcal{L}_\bullet \to F$, where the sheaf $\mathcal{L}_i$ is isomorphic to $(\mathcal{O}_Y(-n_i))^{k_i}$ with $n_i > 0$. By Proposition 4.2.2, the direct images $p_*(\mathcal{L}_i)$ are zero and the direct images $\mathcal{R}_i = R^1 p_*(\mathcal{L}_i)$ are locally-free sheaves. We deduce that the direct images of F are given by

$$R^j p_*(F) = \mathcal{H}_{1-j}(\mathcal{R}_\bullet).$$

In particular, these direct images vanish for $j \geq 2$.

Corollary 4.2.8 *Let X be a projective algebraic curve, F a coherent algebraic sheaf on $Y = S \times X$. Then, for all points $s \in S$,*

$$R^1 p_*(F)_s \otimes_{\mathcal{O}_s} \mathbb{C} \simeq H^1(X, F(s)).$$

Proof With the notation used in the example above the left hand side is the cokernel of the map

$$H^1(X, \mathcal{L}_1(s)) \longrightarrow H^1(X, \mathcal{L}_0(s)).$$

By virtue of the exact sequence $\mathcal{L}_1(s) \to \mathcal{L}_0(s) \to F(s) \to 0$, we see that the cokernel can be identified with $H^1(X, F(s))$. □

Corollary 4.2.9 *Suppose that the sheaf F is S-flat. For all $s \in S$, the canonical map*

$$p_*(F)_s \otimes_{\mathcal{O}_s} \mathbb{C} \longrightarrow H^0(X, F(s))$$

has kernel $\mathrm{Tor}_2^{\mathcal{O}_s}(R^1p_*(F)_s, \mathbb{C})$ *and cokernel* $\mathrm{Tor}_1^{\mathcal{O}_s}(R^1p_*(F)_s, \mathbb{C})$. *Moreover, for $i \geq 1$, we have isomorphisms*

$$\mathrm{Tor}_{i+2}^{\mathcal{O}_s}(R^1p_*(F)_s, \mathbb{C}) \simeq \mathrm{Tor}_i^{\mathcal{O}_s}(p_*(F)_s, \mathbb{C}).$$

Proof Consider the resolution already seen above $\mathcal{L}_\bullet \to F$ in a neighbourhood of the fibre over s. Let $N = \ker(\mathcal{L}_0 \to F)$. Then the sheaf $\mathcal{A} = R^1p_*(N)$ has a locally-free resolution

$$\cdots \longrightarrow \mathcal{R}_2 \longrightarrow \mathcal{R}_1 \longrightarrow \mathcal{A} \longrightarrow 0$$

and this sequence remains exact after tensoring by $\mathbb{C}$ over $\mathcal{O}_s$ because of flatness. This is the sequence given by taking cohomology of the resolution of $N(s)$. Hence, the sheaf $\mathcal{A}$ is locally-free and so we have obtained a map of locally-free sheaves over a neighbourhood of s with kernel $p_*(F)$ and cokernel $R^1p_*(F)$. The 0-sequence of complexes (exact up to quasi-isomorphism)

$$\begin{array}{ccccccccc} & & & & 0 & & & & \\ & & & & \downarrow & & & & \\ 0 & \longrightarrow & p_*(F) & \longrightarrow & \mathcal{A} & & & & \\ & & & & \downarrow & & & & \\ & & & & \mathcal{R}_0 & \longrightarrow & R^1p_*(F) & \longrightarrow & 0 \\ & & & & \downarrow & & & & \\ & & & & 0 & & & & \end{array}$$

gives the exact sequence we need after tensoring with $\mathbb{C}$: indeed, the linear map $\mathcal{A}(s) \to \mathcal{R}_0(s)$ has kernel $H^0(X, F(s))$ and cokernel $H^1(X, F(s))$. □

Example.
We consider the family of all sheaves F on $\mathbf{P}_1$ given as the cokernel of a non-zero map $s : \mathcal{O} \to \mathcal{O}(1)^2$. The family of all such maps s can be identified with $S = \mathbb{C}^4 \setminus \{0\}$ and we get a "universal" morphism

$$\sigma : \mathcal{O}_{S \times \mathbf{P}_1} \longrightarrow q^*(\mathcal{O}_{\mathbf{P}_1}(1)^2)$$

over $S \times \mathbf{P}_1$, where q denotes the projection to the second factor. The cokernel F of σ is clearly S-flat. Over the point defined by s, the cokernel $F(s)$ can be identified with $\mathcal{O}(2)$ unless this map vanishes at a point $x \in \mathbf{P}_1$ in which case we obtain $F(s) = \mathcal{O}(1) \oplus \mathbb{C}_x$. For the module defined by $G = F(-3)$ we then have $p_*(G) = 0$ even although $H^0(\mathbf{P}_1, G(s)) = \mathbb{C}$ when the map s vanishes. Note that in this case the dimension of $H^1(\mathbf{P}_1, F(s))$ is not constant so the sheaf $R^1p_*(F)$ is not locally-free in view of Corollary 4.2.8.

Example.
Here is an analogous example but where we also have that the module F is locally-free. We consider all vector bundles F over $\mathbf{P}_1$ given as an extension

$$0 \longrightarrow \mathcal{O}(-2) \longrightarrow F \longrightarrow \mathcal{O} \longrightarrow 0;$$

in other words, this sequence is exact. These extensions are classified by $H^1(\mathbf{P}_1, \mathcal{O}(-2)) = \mathbb{C}$ (cf. chapter 7): we associate to such an extension the image ω of 1 under the coboundary map in the cohomology long exact sequence. Consider the family of all these extensions which gives a bundle F over $\mathbb{C} \times \mathbf{P}_1$. Then we have an exact sequence of sheaves over $\mathbb{C}$:

$$0 \longrightarrow p_*(F) \longrightarrow \mathcal{O} \longrightarrow \mathcal{O} \otimes H^1(\mathbf{P}_1, \mathcal{O}(-2)) \longrightarrow R^1p_*(F) \longrightarrow 0.$$

It follows that the direct image $p_*(F)$ vanishes although over the fibre above 0 we have $H^0(\mathbf{P}_1, F(0)) = \mathbb{C}$.

Corollary 4.2.10 *If the sheaf F is S-flat then the Euler-Poincaré characteristic $s \mapsto \chi(s)$ is locally constant.*

Indeed, this number is just $\mathrm{rk}(\mathcal{R}_0) - \mathrm{rk}(\mathcal{A})$.

Corollary 4.2.11 *Suppose that the sheaf F is S-flat. Then the functions $s \mapsto h^0(F(s))$ and $s \mapsto h^1(F(s))$ are bounded and lower semi-continuous. If these functions are constant and if S is integral then the sheaves $p_*(F)$ and $R^1p_*(F)$ are locally-free.*

Classically, this corollary is called the Semi-continuity Theorem. It follows immediately from Corollaries 4.2.8 and 4.2.9. Even if we do not know that the variety is reduced, sometimes we can still deduce that the direct images are locally-free using the following result:

Corollary 4.2.12 *Suppose that the sheaf F is S-flat. Let $s \in S$ be a point such that $H^0(X, F(s)) = 0$. Then the sheaf $R^1p_*(F)$ is locally-free in a neighbourhood of the point s. Moreover, if $H^1(X, F(s)) = 0$ then the direct image $p_*(F)$ is locally-free in a neighbourhood of s.*

The proof follows from Corollary 4.2.9 or more directly from the description of the direct image sheaves $p_*(F)$ and $R^1p_*(F)$ as the cohomology sheaves of the complex $\mathcal{R}_\bullet$ given above.

Corollary 4.2.13 *Let F be a coherent algebraic sheaf on $S \times X$ and $s \in S$. For m large enough, the canonical linear map*

$$p_*(F(m))_s \otimes_{\mathcal{O}_s} \mathbb{C} \longrightarrow H^0(X, F(m)(s))$$

is an isomorphism.

Proof The difficulty lies in the fact that we do not have any flatness hypothesis on F, which prohibits us from using Corollary 4.2.9. For all coherent algebraic sheaves G on S we have a canonical map

$$\varphi_{m,G} : p_*(F(m)) \otimes G \longrightarrow p_*(F(m) \otimes p^*(G))$$

which is obviously an isomorphism if G is locally-free. If we choose a presentation of G by locally-free sheaves $L_1 \to L_0 \to G \to 0$ we obtain, by Theorem 4.2.7, the following commuting diagram of exact sequences:

$$\begin{array}{ccccccc} p_*(F(m)) \otimes L_1 & \to & p_*(F(m)) \otimes L_0 & \to & p_*(F(m)) \otimes G & \to 0 \\ \downarrow & & \downarrow & & \downarrow & \\ p_*(F(m) \otimes p^*(L_1)) & \to & p_*(F(m) \otimes p^*(L_0)) & \to & p_*(F(m) \otimes p^*(G)) & \to 0 \end{array}$$

and it follows that the map $\varphi_{m,G}$ is an isomorphism for $m \geq m_0$ (where m_0 depends on G). If we take G to be the structure sheaf of a point s then we obtain the desired result. □

4.3 The Hilbert polynomial

Let X be a projective algebraic curve equipped with an embedding into projective space. We call the degree of the invertible sheaf $\mathcal{O}_X(1)$ the *degree* of X; it is also the number of intersection points of the curve with a general hyperplane. This number will be denoted by $\deg(X)$.

Consider a coherent algebraic sheaf F of rank r and Euler-Poincaré characteristic χ on the curve X. The *Hilbert polynomial* P_F of F is defined by

$$P_F(n) = \chi(F(n)) = rn \deg(X) + \chi.$$

Now, consider another algebraic variety S and a coherent algebraic sheaf F on $X \times S$. It is remarkable that the flatness of F over S can be written in terms of the Hilbert polynomial, at least if the variety S is integral.

Theorem 4.3.1 *If the variety S is integral then the sheaf F is S-flat if and only if the coefficients of the Hilbert polynomial of $F(s)$ remain constant over S.*

Proof We have already seen in Corollary 4.2.10 that the flatness assumption implies that the Euler-Poincaré characteristic of $F(m)(s)$ remains constant as s varies. We will prove the converse. Suppose now that S is integral. On $Y = S \times X$, consider an exact sequence $0 \to N \to L \to F \to 0$ of coherent algebraic sheaves with L locally-free. After tensoring by $\mathbb{C}$ over the point s, we obtain an exact sequence

$$0 \longrightarrow \mathcal{T}or_1^{\mathcal{O}_s}(F, \mathbb{C}) \longrightarrow N(s) \longrightarrow L(s) \longrightarrow F(s) \longrightarrow 0$$

of coherent algebraic sheaves on the curve X. On the other hand, the variety S is integral and so it follows from Corollary 4.2.10 that, for m large enough, the sheaf $p_*(F(m))$ is locally-free. By Theorem 4.2.7, we have the exact sequence $0 \to p_*(N(m)) \to p_*(L(m)) \to p_*(F(m)) \to 0$ and this sequence remains exact after tensoring by $\mathbb{C}$. Taking Serre's Theorem B into account, the second horizontal map of the commuting diagram

$$\begin{array}{ccc} p_*(N(m)) \otimes_{\mathcal{O}_s} \mathbb{C} & \longrightarrow & p_*(L(m)) \otimes_{\mathcal{O}_s} \mathbb{C} \\ \downarrow & & \downarrow \\ H^0(X, N(m)(s)) & \longrightarrow & H^0(X, L(m)(s)) \end{array}$$

has kernel $H^0\big(X, \mathcal{T}or_1^{\mathcal{O}_s}(F(m), \mathbb{C})\big)$ and we have seen above that the first map injects. Now, the vertical maps are isomorphisms by Corollary 4.2.13. Hence, $H^0\big(X, \mathcal{T}or_1^{\mathcal{O}_s}(F(m), \mathbb{C})\big)$ is zero for large enough m. It follows that the Hilbert polynomial of the coherent algebraic sheaf $\mathcal{T}or_1^{\mathcal{O}_s}(F, \mathbb{C})$ is identically zero. So this sheaf is zero and thus the sheaf F is flat over S. $\square$

Theorem 4.3.2 *Let F be a coherent algebraic sheaf on $S \times X$. If the direct image $p_*(F(m))$ is locally-free for m large enough then F is S-flat.*

This result is still true if S is not integral. The only point in the preceding proof where we use this hypothesis is in showing that the sheaf $p_*(F(m))$ is locally-free. Then the result follows from the previous one.

4.4 Quotients of a given bundle

We are going to study how to put a structure of an algebraic variety on the set of coherent quotient modules of a fixed Hilbert polynomial P of a given locally-free sheaf E. We start with some lemmas.

Lemma 4.4.1 (Grothendieck) *Let E be a vector bundle on the projective line. Then E is a direct sum of invertible bundles.*

Proof We induct on the rank r of E. If i is a very large integer then the Serre Duality Theorem and Theorem B imply that $E(-i)$ has no non-zero sections. Consider the largest integer i such that the bundle $E(-i)$ admits a non-zero section. This section can never vanish, so that we obtain an exact sequence of vector bundles

$$0 \longrightarrow \mathcal{O}(i) \longrightarrow E \longrightarrow F \longrightarrow 0,$$

where F is a bundle of rank $r-1$. Then we have $F = \bigoplus_j \mathcal{O}(j)^{r_j}$ with $j \leq i$ if $r_j \neq 0$ because otherwise $E(-i-1)$ would have a non-zero section. Hence, we have $H^1\big(X, \mathcal{H}om(F, \mathcal{O}(i))\big) = 0$ which implies that the exact sequence above splits. □

Lemma 4.4.2 *Let $f : X \to Y$ be a finite morphism of algebraic varieties and let F be a coherent algebraic sheaf on X.*

(a) The sheaf $f_*(F)$ is coherent and $R^q f_*(F) = 0$ if $q > 0$.

(b) We have $H^q(Y, f_*(F)) = H^q(X, F)$ for all integers q. In particular, if X and Y are projective varieties

$$\chi(X, F) = \chi(Y, f_*(F)).$$

Proof (a) The coherence of $f_*(F)$ follows immediately from Theorem 4.2.5. We show that $R^q f_*(F) = 0$ if $q > 0$. Let $F \to R^\bullet$ be an injective resolution of F. For all affine open subsets V of Y, the complex $\Gamma(V, f_*(R^\bullet))$ is exact in degree bigger than 0 by Serre's Theorem

B applied to the coherent algebraic sheaf F over the affine open set $f^{-1}(V)$. It follows that the map of complexes $f_*(F) \to f_*(R^\bullet)$ is a quasi-isomorphism. and so the higher direct images are zero.

(b) If $F \to R^\bullet$ is an injective resolution of F then the complex $f_*(R^\bullet)$ is a flasque resolution of $f_*(F)$ and so can be used to compute the cohomology of $f_*(F)$:

$$\begin{aligned} H^q(Y, f_*(F)) &= H^q(\Gamma(Y, f_*(R^\bullet))) \\ &= H^q(\Gamma(X, R^\bullet)) = H^q(X, F) \end{aligned}$$

which completes the proof. □

Lemma 4.4.3 *Let $\varphi : E \to F$ be a map of vector bundles of rank n and m respectively on an algebraic variety S. We suppose that the rank of $\varphi(s)$ is at least $m - r$ at each point s of S. We let $\mathcal{G}$ denote the coherent cokernel sheaf of φ.*

Let Z^r be the sub-variety of S defined locally by the minors of order $m - r + 1$ of φ. A morphism $f : S' \to S$ factors through to Z^r if and only if the sheaf $f^*(\mathcal{G})$ is locally-free of rank r.

Proof We use without proof the well known and elementary fact that in the open set Ω of the vector space of m by n matrices of rank $\geq m - r$, the sub-variety defined by the minors of order $m - r + 1$ is smooth. If the morphism f factorizes through to Z^r then the sheaf $f^*(\mathcal{G})$ is locally-free of rank r; indeed, the assertion is local and we can assume that the variety S is integral. But then the result follows from Proposition 1.7.2.

For the other way round, if $\mathcal{G}' = f^*(\mathcal{G})$ is locally-free of rank r, we consider the inverse images E' and F' of the bundles E and F under the morphism f; we write $F' = K \oplus L$ locally, where L is a sub-bundle of rank r and K is the inverse image of 0 under the projection $F' \to \mathcal{G}'$. The matrix of $\varphi' = f^*(\varphi)$ can be written

$$\varphi' = \begin{pmatrix} a \\ 0 \end{pmatrix}.$$

It follows that the ideal of $\mathcal{O}_{S'}$ given by the minors of order $m - r + 1$ of φ' is zero. This means that the morphism f factorizes through to Z^r. □

Now fix a locally-free sheaf E on a connected smooth projective curve X. Consider the coherent quotient $\mathcal{O}_X$-modules $G = E/F$ of E of a given Hilbert polynomial P. We set $A_k = \Gamma(X, \mathcal{O}_X(k))$. There exists an integer ν such that $H^1(X, E(i)) = 0$ for $i \geq \nu$.

Lemma 4.4.4 *We can choose the integer ν so that for all quotient modules $G = E/F$ of Hilbert polynomial P and for all integers $k \geq 0$, we have*

(a) $H^1(X, F(\nu + k)) = 0$ and
(b) $A_k H^0(X, F(\nu)) = H^0(X, F(\nu + k))$.

The latter equation means that the vector space of sections of $F(\nu + k)$ is generated by the products of sections of $\mathcal{O}_X(k)$ and of $F(\nu)$.

Proof By choosing a centre of projection which does not lie on X, we obtain a finite morphism $X \to \mathbf{P}_1$ such that $f^*(\mathcal{O}_{\mathbf{P}_1}(1)) = \mathcal{O}_X(1)$. Then $f_*(F)$ is a coherent sub-module of $f_*(E)$ which is a torsion-free coherent sheaf and, hence, locally-free (cf. exercises, chapter 2); this also holds for $f_*(F)$. So, by Lemma 4.4.1,

$$f_*(F) = \bigoplus_j \mathcal{O}(j)^{r_j}.$$

Since the Hilbert polynomial of $f_*(F)$ coincides with that of F by Lemma 4.4.2, the sum $\sum_{j, r_j \neq 0} j r_j$ is fixed and the set of integers j we obtain is bounded above as F varies. It follows that this family is finite. Part (a) follows from the fact that $f_*(F(i)) = f_*(F)(i)$ and from Lemma 4.4.2. As far as part (b) is concerned, it suffices to check it when $X = \mathbf{P}_1$ and $F = \mathcal{O}_{\mathbf{P}_1}(i)$. But then $\bigoplus_n H^0(\mathcal{O}_{\mathbf{P}_1}(n))$ is the algebra of homogeneous polynomials in two variables and the canonical map

$$A_k \otimes H^0(\mathcal{O}_{\mathbf{P}_1}(\nu + i)) \longrightarrow H^0(\mathcal{O}_{\mathbf{P}_1}(\nu + i + k))$$

is obviously surjective provided we choose $\nu \geq -i$. □

Let P be a degree 1 polynomial. Let $\mathbf{Hilb}^P(E)$ denote the set of coherent sheaves E/F with Hilbert polynomial P which are quotients of E by a coherent sub-sheaf F. Our aim is to give $\mathbf{Hilb}^P(E)$ the natural structure of a projective variety.

For any algebraic variety S, we let $E_S = \mathcal{O}_S \boxtimes E$ denote the inverse image of E under the second projection $q : S \times X \to X$ and we let $\underline{\mathbf{Hilb}}^P(E; S)$ denote the set of coherent quotient modules $G = E_S/F$ which are flat over S and such that for each point s of S, the Hilbert polynomial of $G(s)$ is P. If $f : S' \to S$ is a morphism of algebraic varieties then, for all $G \in \underline{\mathbf{Hilb}}^P(E; S)$, the inverse image $(f \times \mathrm{id}_X)^*(G)$

is a coherent module on $S' \times X$ which is S'-flat and which defines a map $G \mapsto (f \times \mathrm{id}_X)^*(G)$ from $\underline{\mathbf{Hilb}}^P(E;S)$ to $\underline{\mathbf{Hilb}}^P(E;S')$. In this way, we obtain a contravariant functor on the category of algebraic varieties.

The following result is due to Grothendieck [29].

Theorem 4.4.5 *The functor $\underline{\mathbf{Hilb}}^P(E;S)$ is representable by a projective algebraic variety $\mathbf{Hilb}^P(E)$.*

If we set $H = \mathbf{Hilb}^P(E)$ then this theorem means that there exists a coherent quotient module $\mathcal{G} \in \underline{\mathbf{Hilb}}^P(E;H)$ on $H \times X$ which induces a functorial isomorphism $f \mapsto (f \times \mathrm{id}_X)^*(\mathcal{G})$:

$$\mathrm{Mor}(S, \mathbf{Hilb}^P(E)) \simeq \underline{\mathbf{Hilb}}^P(E;S).$$

In particular, the underlying set of this projective variety is just the set of coherent quotient sheaves of E with Hilbert polynomial P.

Let ν be an integer such that $H^1(X, E(i)) = 0$ for $i \geq \nu$ and satisfying the conditions of Lemma 4.4.4. Then we have the following constraints for each quotient $G = E/F \in \mathbf{Hilb}^P(E;S)$ and each integer $m \geq 0$:

$$(\mathrm{H}_1)\ H^1(F(\nu+m)) = 0,$$
$$(\mathrm{H}_2)\ A_m H^0(F(\nu)) = H^0(F(\nu+m)).$$

We let $\mathbf{Grass}^{P(\nu)}(H^0(E(\nu)))$ denote the Grassmannian of vector subspaces of codimension $P(\nu)$ in $H^0(E(\nu))$. Consider the map

$$\Phi : \mathbf{Hilb}^P(E) \longrightarrow \mathbf{Grass}^{P(\nu)}(H^0(E(\nu)))$$

which associates to a quotient module $G = E/F$ the sub-space $H^0(F(\nu))$ of $H^0(E(\nu))$. This vector sub-space certainly has codimension $P(\nu)$ because of (H_1). The map Φ is injective: indeed, it follows from condition (H_2) that the sheaf $F(\nu)$ is generated by $H^0(F(\nu))$ and so F can be recovered from the sub-space $H^0(F(\nu))$ as the image of the composite

$$H^0(F(\nu)) \otimes \mathcal{O}_X(-\nu) \longrightarrow E$$

induced by evaluation.

Problem: What is the image of Φ?

To each vector sub-space $\Gamma \in \mathbf{Grass}^{P(\nu)}(H^0(E(\nu)))$ we can associate the cokernel $G = E/F$ of the canonical map $\Gamma \otimes \mathcal{O}_X(-\nu) \to E$.

Lemma 4.4.6 *The map Φ is injective and its image is the set Σ of vector sub-spaces $\Gamma \in \mathbf{Grass}^{P(\nu)}(H^0(E(\nu)))$ which satisfy the condition*

$$\dim(H^0(E(m))/\Gamma A_{m-\nu}) = P(m) \text{ if } m \geq \nu.$$

Proof Obviously, conditions (H_1) and (H_2) imply that we have an exact sequence

$$H^0(F(\nu)) \otimes A_{m-\nu} \longrightarrow H^0(E(m)) \longrightarrow H^0(G(m)) \longrightarrow 0$$

for each $m \geq \nu$ and it follows that the point defined by the sub-space $H^0(F(\nu))$ belongs to Σ.

Conversely, if $\Gamma \in \Sigma$ then, for the associated sheaf G, we have the exact sequence

$$\Gamma \otimes A_{m-\nu} \longrightarrow H^0(E(m)) \longrightarrow H^0(G(m)) \longrightarrow 0$$

for m large enough, which proves that the associated module G has Hilbert polynomial $P_G = P$. Then we have obtained a map $\Psi : \Sigma \to \mathbf{Hilb}^P(E)$ such that $\Psi \circ \Phi = \mathrm{id}_{\mathbf{Hilb}^P(E)}$. We will show that $\Phi \circ \Psi = \mathrm{id}_\Sigma$. We have the inclusions

$$\Gamma \subset H^0(F(\nu)) \subset H^0(E(\nu))$$

and, since Γ and $H^0(F(\nu))$ have codimension $P(\nu)$, they must coincide. Hence, the image of the map Φ is exactly Σ. □

The variety $\mathbf{Hilb}^P(E)$

We must now show that Σ can be given the natural structure of an algebraic variety.

For $\Gamma \in \mathbf{Grass}^{P(\nu)}(H^0(E(\nu)))$, consider the family of maps

$$\Gamma \otimes \mathcal{O}_X(-\nu) \longrightarrow E$$

parametrized by the Grassmannian $S = \mathbf{Grass}^{P(\nu)}(H^0(E(\nu)))$. Let $\mathcal{U}$ denote the universal sub-bundle of the trivial bundle with fibre $H^0(E(\nu))$ over the Grassmannian S. Over $S \times X$, we have a canonical map of locally-free sheaves

$$\mathcal{U} \boxtimes \mathcal{O}_X(-\nu) \longrightarrow \mathcal{O}_S \boxtimes E,$$

and we denote its cokernel by $\mathcal{G}$. On the other hand, if we let $p : S \times X \to S$ denote the first projection then, by the choice of ν, we already know that $R^1 p_*(\mathcal{G}(m)) = 0$ for $m \geq \nu$.

Lemma 4.4.7 *Let $S' \subset S$ be an algebraic sub-variety of S. There exists an integer $\nu'' \geq \nu$ such that for all $m \geq \nu''$ the canonical map*

$$p_*(\mathcal{G}(m)) \otimes \mathcal{O}_{S'} \longrightarrow p_*(\mathcal{G}(m)|_{S'})$$

is an isomorphism.

This statement can be obtained in the same way as in the proof of Corollary 4.2.13.

Consider the map of locally-free sheaves on S

$$\varphi_m : \mathcal{U} \otimes H^0(\mathcal{O}_X(m-\nu)) \longrightarrow \mathcal{O}_S \otimes H^0(E(m)).$$

Applying Serre's Theorem B for direct images, we can find an integer $\nu' \geq \nu$ such that for $m \geq \nu'$, the cokernel of φ_m can be identified with $p_*(\mathcal{G}(m))$. One such integer will be fixed by the following.

Lemma 4.4.8 *There exists a finite number of locally closed sub-varieties $S_j \subset S$ which cover S such that the sheaves $\mathcal{G}|_{S_j}$ are flat over S_j.*

Proof It suffices to find sub-varieties S_j so that for each $m \geq \nu$, the sheaf $p_*(\mathcal{G}(m))|_{S_j}$ is locally-free. The question is local on S: we can replace it with a neighbourhood of $s \in S$. There exists $\xi \in H^0(\mathcal{O}_X(1))$ such that the map given by multiplication by ξ,

$$\xi : \mathcal{G} \longrightarrow \mathcal{G}(1),$$

injects over a neighbourhood $U \times X$ of $\{s\} \times X$. Indeed, the ξ which are not suitable are contained in a finite union of hyperplanes of $H^0(\mathcal{O}_X(1))$. Let $\mathcal{Q}$ denote the cokernel of such a map. This is a torsion sheaf supported on the sub-variety V_ξ given by the equation $\xi = 0$ and we have an isomorphism $\mathcal{Q} \simeq \mathcal{Q}(1)$ given by the choice of a section $\eta \in H^0(\mathcal{O}_X(1))$ which does not vanish on V_ξ. For $m \geq \nu$ we then have the exact sequences of coherent sheaves on U

$$0 \longrightarrow p_*(\mathcal{G}(m)) \longrightarrow p_*(\mathcal{G}(m+1)) \longrightarrow p_*(\mathcal{Q}) \longrightarrow 0.$$

So it suffices to choose the sub-varieties S_j so that the sheaves $p_*(\mathcal{G}(\nu))|_{S_j}$ and $p_*(\mathcal{Q})|_{S_j}$ are locally-free. Then it follows from Lemma 4.4.7 above that $p_*(\mathcal{G}(m)|_{S_j})$ is locally-free for m large enough. Theorem 4.3.2 implies that the sheaf $\mathcal{G}|_{S_j}$ is flat on S_j. □

We can now apply what we have seen in section 4.2: the Hilbert polynomial of the cokernel $\mathcal{G}(s)$ is constant on S_j. Let Q_j be this polynomial. For $m \geq \nu$ and $s \in S_j$,

$$\dim \operatorname{coker} \varphi_m(s) \geq Q_j(m),$$

and we have equality when $m \geq \nu'$.

Consider the open subset U_m of S of points s such that

$$\dim \operatorname{coker} \varphi_m(s) \leq P(m).$$

On the other hand, consider the determinantal sub-variety Σ_m of U_m of points where the dimension of the cokernel of $\varphi_m(s)$ is exactly $P(m)$, in other words, the sub-variety defined by the minors of order $R(m)+1$, where R is the polynomial $P_E - P$.

Lemma 4.4.9 *There exists an integer $\ell \geq \nu'$ such that*

(i) for $m \geq \ell$,

$$\Sigma = \bigcap_{\nu \leq i \leq m} \Sigma_i,$$

(ii) the intersection

$$U_\infty = \bigcap_{m \geq \nu} U_m$$

is open in S.

Proof We choose ℓ large enough so that for $m \geq \ell$ the evaluation map at m separates the polynomials $\{P, Q_1, \ldots, Q_k\}$ and gives rise to an increasing function on this set of polynomials. Then we have, for $m \geq \ell$,

$$\Sigma_m = \bigcup_{j, Q_j = P} S_j.$$

So $\Sigma_m = \Sigma_{m+1}$ for $m \geq \ell$. Obviously, by definition, Σ is the intersection of the Σ_m: this implies part (i). Similarly, we see that

$$U_m = \bigcup_{j, Q_j \leq P} S_j$$

for $m \geq \ell$, and the right hand side is independent of m. This implies the second part. □

Consider the closed sub-scheme of U_∞ defined by $\bigcap_{m \geq \nu} \Sigma_m$. Let H denote this sub-scheme which has Σ as its underlying set. For m large enough, we have an isomorphism

$$p_*(\mathcal{G}(m))|_H \simeq p_*(\mathcal{G}(m)|_H)$$

and the left hand side is a locally-free sheaf by Lemma 4.4.3. It follows that the sheaf $\mathcal{G}|_H$ is flat over H and for $s \in H$, the Hilbert polynomial of $\mathcal{G}(s)$ is P.

The universal property

Let S' be an algebraic variety, $\mathcal{G}' = E_{S'}/\mathcal{F}'$ a quotient module in $\underline{\mathbf{Hilb}}^P(E; S')$ and $p' : S' \times X \to S'$ the projection to the first factor. By the choice of ν, the sheaf $p'_*(\mathcal{G}'(\nu))$ is, from Corollary 4.2.12, a locally-free quotient sheaf of the trivial bundle on S' with fibre $H^0(E(\nu))$; it defines a map

$$f : S' \longrightarrow \mathbf{Grass}^{P(\nu)}(H^0(E(\nu)))$$

by the universal property of the Grassmannian. On the other hand, from the flatness of $\mathcal{G}'$ and the choice of ν we have an exact sequence of locally-free sheaves

$$p'_*(\mathcal{F}(\nu)) \otimes A_{m-\nu} \longrightarrow \mathcal{O}_{S'} \otimes H^0(E(m)) \longrightarrow p'_*(\mathcal{G}'(m)) \longrightarrow 0$$

for $m \geq \nu$. Moreover, the sheaf $p'_*(\mathcal{G}'(m))$ is locally-free of rank $P(m)$. It follows that the map f factors through the sub-scheme H. From the definition of the map f the sheaf $p'_*(\mathcal{F}'(m))$ is the inverse image of the universal sub-bundle $\mathcal{U}$ under the base change $f : S' \to S$. The sheaf $\mathcal{G}'$ is exactly the cokernel of the bundle map

$$p'_*(\mathcal{F}'(\nu)) \boxtimes \mathcal{O}_X(-\nu) \longrightarrow \mathcal{O}_{S'} \boxtimes E.$$

This map is obtained from ϵ by the base change $f : S' \to S$. It follows that the cokernel is the inverse image of $\mathcal{G}$ under f. Hence, we have shown that the functorial morphism

$$\mathrm{Mor}(S', \mathbf{Hilb}^P(E)) \longrightarrow \underline{\mathbf{Hilb}}^P(E; S')$$

given by the the universal family $\mathcal{G}|_H$ surjects. This map is also injective because given a map $f : S' \to H$, if we set $\mathcal{G}' = (f \times \mathrm{id}_X)^*(\mathcal{G})$ then the map f is determined by the quotient sheaf $p'_*(\mathcal{G}'(\nu))$ and so by $\mathcal{G}'$. □

Corollary 4.4.10 *Suppose that ν is an integer satisfying the conditions of Lemma 4.4.4. Then the canonical map*

$$i : \mathbf{Hilb}^P(E) \longrightarrow \mathbf{Grass}^{P(\nu)}(H^0(E(\nu)))$$

which associates the sub-space $H^0(F(\nu))$ of codimension $P(\nu)$ in $H^0(E(\nu))$ to the quotient module $G = E/F$ is an embedding.

Indeed, this follows from the construction we have given of the variety $\mathbf{Hilb}^P(E)$.

The variety $\mathbf{Hilb}^P(E)$ is projective

We use the criterion of properness: we must show that if (C, a) is a smooth affine curve with a given point a, then all maps $f : C' = C \setminus \{a\} \to \mathbf{Hilb}^P(E)$ extend to C.

To such a map f there corresponds a coherent quotient sheaf $\mathcal{G}' = E_{C'}/\mathcal{F}'$ in $\underline{\mathbf{Hilb}}^P(E; C')$. The sheaf $\mathcal{F}'$ extends over $C \times X$ to give a coherent sub-sheaf $\mathcal{F}$. Indeed, $\mathcal{F}'$ is a quotient of a locally-free sheaf $\mathcal{L}' = \mathcal{O}_{C'} \boxtimes \mathcal{O}_X(-k)$ for large enough k. Let z be a parameter in a neighbourhood of a. This gives us a map $u : \mathcal{L}' \to E_{C'}$ and the map $z^\ell u$ extends to $C \times X$ to give a map

$$v : \mathcal{L} = \mathcal{O}_C \boxtimes \mathcal{O}_X(-k) \longrightarrow E_C$$

for large enough ℓ. The image of v is a sub-module of finite type (hence coherent) which extends $\mathcal{F}'$. Since every union of an increasing sequence of sub-modules of finite type is still of finite type, there exists a maximal sub-module of finite type E_C which coincides with $\mathcal{F}'$ over the open set $C' \times X$. Two such maximal sub-modules coincide: because if $\mathcal{F}_1$ and $\mathcal{F}_2$ are two such sub-modules then $\mathcal{F}_1 + \mathcal{F}_2$ is also an extension of $\mathcal{F}'$ of finite type and so $\mathcal{F}_1 = \mathcal{F}_2$. It follows that there exists a largest extension which we denote by $\mathcal{F}$ and which is still of finite type. Let $\mathcal{G} = E_C/\mathcal{F}$. It suffices to show that the coherent module $\mathcal{G}$ is flat over C: indeed, from the invariance of the Hilbert polynomial (Corollary 4.2.10), it has P as its Hilbert polynomial and so this gives us the required extension of the map f.

By Proposition 4.1.2 we need to show that $\mathcal{T}or_1^{\mathcal{O}_{C,a}}(\mathcal{G}, \mathbb{C}) = 0$. Let z be a parameter which vanishes at a in a neighbourhood of a on the curve C. This amounts to proving that the morphism $\mathcal{G} \xrightarrow{z} \mathcal{G}$ induced by z injects. Consider its kernel: its inverse image in E_C gives rise to a coherent sub-module $\overline{\mathcal{F}}$ of E_C containing $\mathcal{F}$ which coincides with $\mathcal{F}$ over the open set $C' \times X$. So it equals $\mathcal{F}$. Hence, the kernel is zero and this gives us flatness. □

4.5 Exercises

Consider the vector bundle $E = \mathcal{O}(1)^2$ on the projective line $\mathbf{P}_1$ and take $P(n) = n + 3$ as the Hilbert polynomial.

4.1 Show that the Hilbert scheme $\mathbf{Hilb}^P(E)$ is isomorphic to the projective space $\mathbf{P}_3$.

4.2 Let $\mathcal{G}$ be the universal quotient sheaf on $\mathbf{Hilb}^P(E) \times \mathbf{P}_1$. Prove that the points $s \in \mathbf{Hilb}^P(E)$ corresponding to the sheaves which are not locally-free are given by the condition

$$\dim H^1(\mathbf{P}_1, \mathcal{G}(s)(-3)) = 1.$$

4.3 Deduce that the points of the Hilbert scheme corresponding to the locally-free sheaves form the complement of a quadric in $\mathbf{P}_3$.

5

Semi-stability

The set of isomorphism classes of algebraic vector bundles of fixed rank and degree cannot be parametrized by an algebraic variety. To get around this problem we introduce the notion of semi-stable bundle. The study of bundles which are not semi-stable can be reduced to studying semi-stable bundles: in fact, each algebraic bundle admits a canonical increasing filtration called the Harder-Narasimhan filtration [30] whose successive quotients are semi-stable. The category of semi-stable bundles of a fixed slope is an abelian category: it was C.S. Seshadri who first observed this remarkable fact [55]. It allows us to define Jordan-Hölder filtrations for each semi-stable bundle: these filtrations are important in order to understand the points of the moduli spaces. The most significant result of the chapter is Theorem 5.5.1, which shows that the family of semi-stable bundles with given rank and Chern class is bounded.

In the following, X will denote a smooth connected projective algebraic curve of genus g.

5.1 Bounded families

Let $\mathcal{V}$ be a set of isomorphism classes of algebraic vector bundles on X. We say that $\mathcal{V}$ is *bounded* if there exist an algebraic variety S (of finite type) and an algebraic vector bundle F on $S \times X$ such that $\mathcal{V}$ is contained in the set of vector bundle classes $F(s)$, as s varies over S. Such a bundle F is also called an algebraic family of vector bundles parametrized by S. If $\mathcal{V}$ is such a bounded set then the set of values taken by the pair (rank, degree) on this set is finite. Moreover, the set of values taken by $h^1(E)$ for $E \in \mathcal{V}$ is finite by Corollary 4.2.11, and the same holds for $h^0(E)$.

Example.
By Theorem 4.4.5, the family of vector bundles of rank r and degree d which are quotients of a given bundle E is bounded. The same does not follow if we only fix the rank and degree.

Proposition 5.1.1 *The set of isomorphism classes of vector bundles of rank r and degree d is not bounded if $r \geq 2$.*

Proof Consider a point $a \in X$ and the vector bundle

$$E_k = \mathcal{O}(-ka) \oplus \mathcal{O}((d+k)a) \oplus \mathcal{O}^{r-2}.$$

Then we have $h^1(E) \geq h^1(\mathcal{O}(-ka)) = k + g - 1$. This family of integers is not bounded and so we have the proposition. □

Note that what stops the family of bundles in this proof from being bounded is the presence of a sub-bundle in the family with arbitrarily large degree.

We start with some points about the structure of coherent modules on a smooth curve; the statements were suggested by the exercises of chapter 2.

5.2 Coherent sheaves on a smooth curve

If F is a coherent sheaf then the dual sheaf $F^* = \mathcal{H}om(F, \mathcal{O})$ of F is also coherent. The kernel T of the canonical map $F \to F^{**}$ is a coherent sub-module which vanishes on the open set where F is locally-free: this is the *torsion* sub-sheaf of F. The quotient $G = F/T$ is then a torsion-free sheaf, in other words, at each point $x \in X$, the module G_x is a torsion-free $\mathcal{O}_{X,x}$-module. Such sheaves are locally-free:

Lemma 5.2.1 *Every torsion-free coherent sheaf G on a smooth curve is locally-free.*

Proof Since the curve X is smooth, the local ring $\mathcal{O}_{X,x}$ is a principal ring and any torsion-free module of finite type over such a ring is free. Hence, for all $x \in X$, the module G_x is free. It follows that G is a locally-free sheaf. □

Lemma 5.2.2 *Every coherent sheaf F is the direct sum of a locally-free sheaf G and a sheaf supported at a finite number of points.*

Proof Let T be the torsion sub-sheaf of F and let $G = F/T$. Then, G is a locally-free sheaf by the preceding lemma, and we have an exact sequence

$$0 \longrightarrow T \longrightarrow F \xrightarrow{p} G \longrightarrow 0.$$

We need to show that this exact sequence splits. We apply the functor $\mathcal{H}om(G, -)$ followed by the sections functor to this sequence to obtain the exact sequence

$$H^0(X, \mathcal{H}om(G, F)) \longrightarrow H^0(X, \mathcal{H}om(G, G)) \longrightarrow H^1(X, \mathcal{H}om(G, T)).$$

The last vector space vanishes since the sheaf $\mathcal{H}om(G, T)$ is supported at a finite number of points. Thus, there exists a map $s : G \to F$ such that $p \circ s = \mathrm{id}_G$. Hence, $F = G \oplus T$. □

5.3 Semi-stability

For a coherent sheaf F of rank $r > 0$ and degree d we call the rational number

$$\mu(F) = \frac{d}{r}$$

the *slope* of F. In what follows, the algebraic vector bundles will be regarded as locally-free sheaves.

Definition 5.3.1 *Let F be an algebraic vector bundle on X of slope μ. We say that F is* semi-stable *if for every non-zero coherent sub-sheaf F' of F, we have $\mu(F') \leq \mu$.*

Note (see chapter 2) that all coherent sub-sheaves F' of F are constrained to be locally-free. Such a sub-sheaf must be contained in a sub-bundle of F of the same rank: indeed, consider the quotient sheaf $F'' = F/F'$ and let T'' denote the torsion sub-sheaf of F''. The quotient $G'' = F''/T''$ is a torsion-free coherent sheaf which must be locally-free on a smooth curve by Lemma 5.2.1. The inverse image G of T'' under the canonical projection $F \to F''$ is a sub-sheaf containing F' and coincides with F' outside a finite number of points. Note that the inclusion is a maximal rank map of locally-free sheaves $F' \to G$ and so this map has non-zero determinant. It follows that $\mu(F') \leq \mu(G)$. Hence, to check that a vector bundle F is semi-stable it suffices to show that, for all non-zero sub-bundles G of F, we have $\mu(G) \leq \mu(F)$.

We can also formulate semi-stability in terms of quotient coherent sheaves: the bundle F is semi-stable if and only if, for every coherent quotient sheaf $F'' = F/F'$ of non-zero rank, we have $\mu(F'') \geq \mu(F)$. This results from the following barycentre formula: let r' and r'' denote the ranks of F' and F'' respectively, and μ' and μ'' their respective slopes; then

$$\mu = \frac{r'\mu' + r''\mu''}{r}.$$

In fact, to prove semi-stability, it suffices to check that the inequality holds for locally-free quotient sheaves.

Definition 5.3.2 *A vector bundle F of slope μ is* stable *if it is semi-stable and has no non-zero sub-bundles of slope μ other than itself.*

Vector bundles of rank 1 are obviously stable. By Grothendieck's Lemma (Lemma 4.4.1), the only semi-stable bundles of rank greater than 1 on the projective line are the bundles $\mathcal{O}^r(k)$. We will show in chapter 8 that on all curves of genus g at least 2 there exist stable bundles of given rank r and degree d. This fails both for $\mathbf{P}_1$ and for elliptic curves (cf. chapter 8).

Proposition 5.3.3 *Let E and F be two semi-stable bundles on the curve X. If* $\mathrm{Hom}(E, F) \neq 0$ *then we have*

$$\mu(E) \leq \mu(F).$$

If, further, E and F are stable and of the same slope then every non-zero map $f : E \to F$ is an isomorphism.

Proof Let $f : E \to F$ be a non-zero map and I the coherent image sheaf of f; this sheaf is locally-free of non-zero rank. Using the definition of semi-stability, we obtain $\mu(E) \leq \mu(I) \leq \mu(F)$ which proves the first part. In the case where E and F are stable of the same slope, the image cannot have rank smaller than that of E or of F. So E and F have the same rank r and the map is of maximal rank except at points. But then the determinant of f, considered as a section of the invertible bundle $\det(F) \otimes \det(E)^{-1}$, must be non-zero everywhere because this bundle has degree 0. Hence, f is an isomorphism. □

Corollary 5.3.4 *Let E be a stable bundle. The only endomorphisms of E are homotheties.*

Proof Let f be an endomorphism of E. Consider the commutative algebra $\mathbb{C}[f]$ generated by f. This algebra is a field by the above proposition. Then it is a finite algebraic extension of $\mathbb{C}$ and so must be $\mathbb{C}$: in other words, f is a homothety. □

Consider a rational number μ and let $\mathcal{C}(\mu)$ denote the category of semi-stable vector bundles of slope μ.

Proposition 5.3.5 *Consider an exact sequence of vector bundles*

$$0 \longrightarrow E \longrightarrow F \longrightarrow G \longrightarrow 0.$$

If E and G are objects of $\mathcal{C}(\mu)$ then so is F.

Proof Let F' be a coherent sub-sheaf of F. Consider the intersection $E' = E \cap F'$ and the image G' of F' in G. If these sheaves are non-zero then we can write $\mu(E') \leq \mu$ and $\mu(G') \leq \mu$. Since we have the exact sequence

$$0 \longrightarrow E' \longrightarrow F' \longrightarrow G' \longrightarrow 0,$$

the slope of F' is the barycentre of $\mu(E')$ and $\mu(G')$ with positive coefficients. From which we have $\mu(F') \leq \mu$. If $E' = 0$ then F' can be identified with a sub-sheaf of G and if $G' = 0$ then it is contained in E'. In either case, the inequality $\mu(F') \leq \mu$ is obvious. □

Proposition 5.3.6 *The category $\mathcal{C}(\mu)$ is abelian.*

Proof This category is obviously an additive sub-category of the category of vector bundles because it has direct sums by the previous proposition. It remains to prove that all morphisms have a kernel and a cokernel. Consider a non-zero map $f : E \to F$ of semi-stable bundles of slope μ. Then the image I of f has slope μ; it is forced to be a sub-bundle of F, so f has constant rank and the kernel and cokernel are vector bundles. We show that the kernel is semi-stable: if not then it would have a sub-bundle of slope strictly greater than μ. But this sub-bundle is also a sub-bundle of E, which contradicts the semi-stability of E. We now show that the cokernel is semi-stable. If not then it has a non-zero locally-free quotient bundle of slope strictly less that μ. This quotient is also a quotient of F, which contradicts the semi-stability of F. □

In this abelian category, descending filtrations are stationary because the rank decreases: we say that the category is *Artinian*. In such a category we have the notion of a Jordan-Hölder filtration.

Jordan-Hölder filtrations

Consider a semi-stable bundle F of slope μ. Such a bundle has a sub-bundle F_1 of slope μ which is stable: if not then we can construct a descending sequence of sub-bundles in $\mathcal{C}(\mu)$ in which the ranks strictly decrease but this is impossible. So F/F_1 is also in $\mathcal{C}(\mu)$ and we can continue the construction. We finally obtain a strictly increasing sequence of vector sub-bundles

$$0 \subset F_1 \subset F_2 \subset \cdots \subset F_k = F$$

in $\mathcal{C}(\mu)$ such that the quotient $\mathrm{gr}_i = F_i/F_{i-1}$ is a stable bundle in $\mathcal{C}(\mu)$. Such a filtration is called a *Jordan-Hölder filtration* of F; the integer k is called the *length* of the filtration and the direct sum $\bigoplus_i \mathrm{gr}_i$ is called the *associated grading.*

Proposition 5.3.7 *Let F be a semi-stable bundle of slope μ. All the Jordan-Hölder filtrations have the same length k. Moreover, if (F_i) and (F_i') are two such Jordan-Hölder filtrations with associated gradings $\bigoplus \mathrm{gr}_i$ and $\bigoplus \mathrm{gr}_i'$ respectively then there exists a permutation $\sigma \in \mathfrak{S}_k$ such that*

$$\mathrm{gr}_i \simeq \mathrm{gr}'_{\sigma(i)}.$$

In particular, the gradings are isomorphic.

Proof Consider two Jordan-Hölder filtrations (F_i) and (F_i') of F of lengths k and k', respectively. Let j be the integer defined by the conditions $F_1 \subset F_j'$ and $F_1 \not\subset F_{j-1}'$. The canonical map $F_1 \to \mathrm{gr}_j'$ is then non-zero and so, by Proposition 5.3.3, it is an isomorphism. Consider the image filtrations $(\overline{F}_i)$ and $(\overline{F}_i')$ in the quotient F/F_1. These still form Jordan-Hölder filtrations with associated gradings

$$\bigoplus_{i\geq 2} \mathrm{gr}_i \quad \text{and} \quad \bigoplus_{i\neq j} \mathrm{gr}_i',$$

respectively. The result now follows by induction on the rank. These filtrations have the same length and so $k = k'$. For the other part, there exists a permutation $\tau : \{2, \ldots, k\} \to \{1, \ldots, k\} \setminus \{j\}$ such that $\mathrm{gr}_i \simeq \mathrm{gr}_{\tau(i)}$ for $i \geq 2$. To obtain σ just extend τ by setting $\sigma(1) = j$. $\square$

5.4 Harder-Narasimhan filtrations

We now show how to classify bundles which are not semi-stable in terms of semi-stable bundles.

Lemma 5.4.1 *Let F be an algebraic vector bundle on the curve X. Then the degrees of its coherent sub-sheaves are bounded above.*

Proof The lemma is obvious on a projective line by Grothendieck's Lemma. To reduce the general case to this we consider a finite morphism $f : X \to \mathbf{P}_1$ such that $f^*(\mathcal{O}_{\mathbf{P}_1}(1)) = \mathcal{O}_X(1)$. So, if F' is a sub-sheaf of F, then the sheaf $G' = f_*(F')$ is a sub-sheaf of a locally-free sheaf $G = f_*(F)$ and, hence, $\chi(G')$ is bounded above. But we have $\chi(F') = \chi(G')$ and so $\chi(F')$ is bounded above. It follows that the degree of F' is bounded above. □

Proposition 5.4.2 *Let F be an algebraic vector bundle on X. Then F has an increasing filtration by vector sub-bundles*

$$0 \subset F_1 \subset F_2 \subset \cdots \subset F_k = F$$

where the grading $\mathrm{gr}_i = F_i/F_{i-1}$ satisfies the following conditions:

(1) the grading gr_i is semi-stable;

(2) $\mu(\mathrm{gr}_i) > \mu(\mathrm{gr}_{i+1})$ for $i = 1, \cdots, k-1$.

This filtration is unique.

The filtration of the proposition is called the *Harder-Narasimhan filtration* of F. Let

$$\mu_{\max}(G) = \mu(\mathrm{gr}_1), \qquad \mu_{\min}(F) = \mu(\mathrm{gr}_k)$$

and put $\mu_i = \mu(\mathrm{gr}_i)$: this is the *Harder-Narasimhan sequence* of F.

Proof If F is already semi-stable then the result is trivial. If not then choose a sub-sheaf F_1 of maximal slope and maximal rank amongst the sub-sheaves of maximal slope. Then F_1 is a semi-stable bundle and the quotient is also locally-free so we can repeat the construction. Since we have $\mu(G) < \mu(F_1)$ for all non-zero sub-sheaves G of F/F_1, we obtain condition (2). For uniqueness, note that if $(F_i)_{i=1,\ldots,k}$ is a filtration which satisfies the stated properties and if F' is a sub-sheaf of F then this filtration induces a filtration (F_i') of F' for which the gradings gr_i' are contained in gr_i. If they are non-zero, we have $\mu(\mathrm{gr}_i') \leq \mu(\mathrm{gr}_i) \leq \mu(\mathrm{gr}_1)$. Hence, we obtain $\mu(F') \leq \mu(\mathrm{gr}_1)$. Moreover, the case of equality forces all of the gr'_i to be zero for $i \geq 2$, in other words, $F' \subset F_1$. It follows that if (G_i) is another filtration satisfying the stated conditions of the proposition then the slopes of F_1 and G_1 are maximal, so they are equal, hence, $F_1 = G_1$. Applying the result to F/F_1 we prove that $F_2 = G_2$ and, continuing in this way, we see that the sequences are equal and so we have uniqueness. □

5.5 The family $\mathcal{S}(r, d)$

Theorem 5.5.1 *The set $\mathcal{S}(r, d)$ of isomorphism classes of semi-stable bundles of given rank r and degree d is bounded.*

Proof Let μ be the slope d/r. If k is an integer such that $\mu + k > 2g - 2$, and L is a degree k line bundle then, by Proposition 5.3.3, we have $\mathrm{Hom}(F \otimes L, \omega_X) = 0$ for every semi-stable bundle F of $\mathcal{S}(r, d)$. By Serre duality, this implies that $H^1(F \otimes L) = 0$. Choose an integer $\nu > 2g - 1 - \mu$ and a point $a \in X$. Then the bundle $F(\nu a)$ is generated by its sections and $H^1(F(\nu a)) = 0$. Put $N = \chi(F(\nu a)) = d + r(\nu + 1 - g)$. Let H be an N-dimensional vector space. For $F \in \mathcal{S}(r, d)$, the vector space $H^0(F(\nu a))$ has dimension N and the choice of an isomorphism $H \simeq H^0(F(\nu a))$ can be composed with the evaluation map to obtain a surjective morphism of vector bundles

$$H \otimes \mathcal{O}(-\nu a) \longrightarrow F.$$

If P is the Hilbert polynomial of the elements of $\mathcal{S}(r, d)$ then this morphism defines a point in the Hilbert scheme $\mathbf{Hilb}^P(H \otimes \mathcal{O}(-\nu a))$. It follows that the family $\mathcal{S}(r, d)$ is bounded. □

Choose the integer ν as above, put $E = H \otimes \mathcal{O}(-\nu a)$ and let Ω denote the set of quotients $F = E/G$ of $\mathbf{Hilb}^P(E)$ which are semi-stable vector bundles and such that the canonical map induced by the projection $H \to H^0(F(\nu a))$ is an isomorphism. We will see later that this set is open. We may also consider the natural action of $\mathrm{SL}(H)$ on $\mathbf{Hilb}^P(E)$. Then Ω is invariant under this action.

Theorem 5.5.2 *The natural map*

$$\Omega/\mathrm{SL}(H) \longrightarrow \mathcal{S}(r, d)$$

is bijective.

Proof It is clear that two points of Ω which are in the same orbit define isomorphic quotients. This defines the map

$$\Omega/\mathrm{SL}(H) \longrightarrow \mathcal{S}(r,d).$$

By the choice of ν, this map is surjective. Furthermore, suppose that two points q and q' define isomorphic quotients F and F'. Given an isomorphism $F \simeq F'$ there exists an automorphism $g \in \mathrm{GL}(H)$ making the diagram

$$\begin{array}{ccc} H & \longrightarrow & H^0(F) \\ {\scriptstyle g}\downarrow & & \downarrow \\ H & \longrightarrow & H^0(F') \end{array}$$

commute. We can suppose, after multiplying by a scalar, that $g \in \mathrm{SL}(H)$. Then the image of q under g is q'. Thus, q and q' are in the same orbit, which gives the injectivity. □

5.6 The family $\mathcal{S}_\alpha(r,d)$

In chapter 7 we will need the following result which generalizes Theorem 5.5.1.

Theorem 5.6.1 *Let $\mathcal{S}_\alpha(r,d)$ be the family of algebraic vector bundles of rank r and degree d on X satisfying the following condition:*

$$\mu(F') \leq \mu + \alpha$$

for all sub-sheaves $F' \subset F$. Then $\mathcal{S}_\alpha(r,d)$ is a bounded family.

Proof It suffices to prove, as in the proof of Theorem 5.5.1, that there exists an integer ν such that, for all $F \in \mathcal{S}_\alpha(r,d)$ and all line bundles L of degree $k \geq \nu - 1$, we have $H^1(F \otimes L) = 0$ because then the bundle $F(\nu a)$ would be generated by its sections and F could be identified with a point of a Hilbert scheme. By Serre duality we must show that, for k large enough, we have $\mathrm{Hom}(F \otimes L, \omega_X) = 0$. But if F'' is a rank 1 locally-free quotient of F with degree d'' then we have $d'' \geq \mu - (r-1)\alpha$. Thus, $\mathrm{Hom}(F \otimes L, \omega_X) = 0$ whenever $k \geq 2g - 1 - \mu + (r-1)\alpha$. □

6

Invariant Geometry[1]

In the preceding chapter we described the set of isomorphism classes of semi-stable vector bundles of given rank and degree as a quotient of a part of a Hilbert scheme, which we saw was open, by the action of a group $\mathrm{SL}(H)$. Here we will show that this quotient can be given the structure of an algebraic variety once we have identified the semi-stable bundles which have the same Jordan-Hölder gradings together. The problem is to determine under which conditions there exists a quotient of an algebraic variety by the action of an algebraic group.

Historically, invariant theory consisted of studying binary forms (i.e. homogeneous polynomials of degree 2) invariant under substitution of variables. For example, consider a quadratic form in two variables $Q(X,Y) = aX^2 + 2bXY + cY^2$. Under the change of variables

$$g : (X,Y) \mapsto (\alpha X + \beta Y, \gamma X + \delta Y),$$

where

$$g = \begin{pmatrix} \alpha & \beta \\ \gamma & \delta \end{pmatrix}$$

is an invertible matrix, the discriminant of the quadratic form $\Delta = ac - b^2$ becomes $\det(g)^2\Delta$; in particular, if $g \in \mathrm{SL}_2$ then the discriminant is invariant.

Question: describe the algebra of polynomial functions in the coefficients of Q which are invariant under substitution by elements of SL_2.

Any such polynomial function φ invariant under SL_2 when we restrict it to the sub-space S of quadratic forms of the form $aX^2 + cY^2$ must be invariant under the substitution $(X,Y) \mapsto (-Y,X)$ and also under $(X,Y) \mapsto (\alpha X, \frac{1}{\alpha}Y)$ for $\alpha \in \mathbb{C}$. Hence, $\varphi|_S$ is a polynomial $P(\Delta)$ in

[1] This chapter was delivered and revised by Christoph Sorger.

$\Delta = ac$. All quadratic forms are conjugate, under a change of variables, to a quadratic form in S and so we see that $\varphi = P(\Delta)$ on the vector space of all quadratic forms. Hence, the algebra that we are looking for is the algebra of polynomials in Δ.

In more variables, the problem can be formulated in the following way: given a vector space V of dimension r, we can make $G = \mathrm{SL}(V)$ act on the vector space $W = S^n(V)$ of homogeneous polynomials of degree n by

$$(g, P) \mapsto P^g = P \circ S^n(g).$$

Consider the polynomial algebra $S(W^*)^G$ of polynomials on W invariant under the action of G. Is this an algebra of finite type? This problem was resolved by Hilbert in 1890 in a surprising way because the proof gives the existence of a finite number of generators for this algebra without making them explicit as we did in the above example.

More generally, if we are given a $\mathbb{C}$-algebra of finite type on which an algebraic group G acts, we can pose the question of understanding the conditions under which the algebra of invariant elements is of finite type. If this happens then the inclusion $A^G \hookrightarrow A$ gives rise to a morphism of algebraic varieties $\operatorname{Spec} A \to \operatorname{Spec} A^G$ and it is the study of this which is called geometric invariant theory.

The most important reference for invariant geometry is the book of D. Mumford and J. Fogarty [47]. One could also consult the book by H. Kraft [34] or the book of P. Newstead [50].

6.1 Quotients

Let $M_n(\mathbb{C})$ be the affine space of matrices of order n. The group GL_n of invertible matrices of order n is given the structure of a natural affine algebraic variety

$$\mathrm{GL}_n = \{g \in M_n(\mathbb{C}) \mid \det g \neq 0\}$$

with coordinate ring $\mathcal{O}(\mathrm{GL}_n) = \mathbb{C}[X_{ij}, \det^{-1}]$.

Definition 6.1.1 *A (Zariski) closed sub-group of GL_n is called an* (affine, linear) algebraic group.

The multiplicative group $\mathbb{C}^*$, the finite groups, the group SL_n of linear maps with determinant 1, the orthogonal groups $\mathrm{O}(q)$ of elements of GL_n which leave invariant a quadratic form q on $\mathbb{C}^n$ are affine algebraic groups. Such a group G is equipped with the structure of a reduced algebraic variety induced by that of GL_n: the multiplication $\mu : G \times G \to G$ (restriction of the multiplication of matrices) is a morphism of algebraic varieties. The map given by inverting elements is an isomorphism of algebraic varieties. Since the variety structure of G is reduced, the open subset of smooth points is non-empty and invariant under translation; it follows that G is smooth. We let G^0 denote the connected component of the identity element.

Definition 6.1.2 *An operation of G on an algebraic variety Y is given by a morphism $\sigma : G \times Y \to Y$ such that*

(1) for all $g \in G$, the morphism $\sigma_g : Y \to Y$ induced by g is an automorphism of Y,

(2) the map $g \mapsto \sigma_g : G \to \mathrm{Aut}(Y)$ from G to the automorphism group of Y is a group homomorphism.

In what follows we fix an operation (we also say an action) σ of an affine algebraic group G on an algebraic variety Y; for $g \in G$ and $y \in Y$, the image $\sigma(g, y)$ will simply be denoted by $g.y$. We define the orbit $\mathrm{Orb}^G(y)$ of a point y of Y under the action of G to be the image of the morphism $g \mapsto g.y$ of G to Y and the stabilizer $\mathrm{Stab}^G(y)$ of y by the action of G to be the closed sub-group of elements of G fixed by y. Let $\psi : G \times Y \to Y \times Y$ be the graph morphism:

$$(g, y) \mapsto (g.y, y).$$

Definition 6.1.3 *Let Y be an algebraic variety equipped with the action of an algebraic group G. The action of G on Y is said to be*

- closed *if all of the orbits of Y under the action of G are closed,*
- proper *if the morphism ψ is proper and*
- free *if the morphism ψ is a closed immersion.*

Definition 6.1.4 *Let Y and Y' be two algebraic varieties equipped with the action of an algebraic group G. A morphism $\varphi : Y \to Y'$ is said to be G-*equivariant *(or G-*invariant*) if the diagram*

$$\begin{array}{ccc} G \times Y & \xrightarrow{\sigma} & Y \\ {\scriptstyle \mathrm{id}_G \times \varphi}\downarrow & & \downarrow{\scriptstyle \varphi} \\ G \times Y' & \xrightarrow{\sigma} & Y' \end{array}$$

commutes.

It follows from the diagram that we have $\phi(g.y) = g.\phi(y)$ for all $g \in G$ and $y \in Y$. Hence, the image of an orbit in Y under ϕ is contained in an orbit in Y'.

Categorical quotients

We assume that we are given an algebraic variety Y equipped with the action of an algebraic group G.

Definition 6.1.5 *A* categorical quotient *of Y by G is defined to be a pair (Z, φ) given by an algebraic variety Z and a G-invariant morphism $\varphi : Y \to Z$ satisfying the following universal property: for all G-invariant morphisms $\varphi' : Y \to Z'$, there exists a unique morphism $\gamma : Z \to Z'$ making the following diagram commute:*

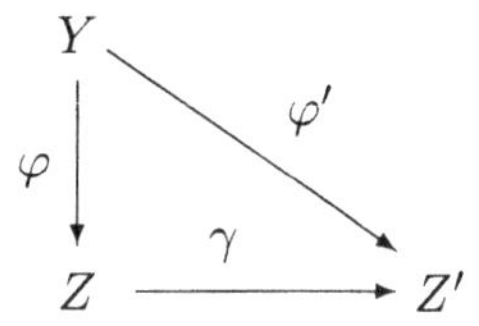

By the universal property, a categorical quotient is unique up to isomorphism.

Example.
Let M_n be the vector space of n by n matrices on which the linear group GL_n acts by conjugation: $(g, a) \mapsto gag^{-1}$ for $g \in \mathrm{GL}_n$ and $a \in M_n$. The coefficients $\sigma_i(a)$ of the characteristic polynomial $t^n - \sigma_1(a)t^{n-1} + \cdots + (-1)^n \sigma_n(a)$ are polynomial functions which are invariant under conjugation and so we obtain a GL_n-equivariant morphism $\varphi : M_n \to \mathbb{C}^n$:

$$a \mapsto (\sigma_1(a), \ldots, \sigma_n(a)).$$

The pair $(\mathbb{C}^n, \varphi)$ is the categorical quotient of M_n by the action of GL_n. Indeed, we can construct a section $s : \mathbb{C}^n \to M_n$ by mapping $(x_1, \ldots, x_n)$

to the matrix

$$\begin{pmatrix} 0 & -1 & 0 & \cdots & \cdots & 0 \\ 0 & 0 & -1 & 0 & \cdots & 0 \\ \vdots & & \ddots & \ddots & \ddots & \vdots \\ \vdots & & & \ddots & \ddots & 0 \\ 0 & \cdots & \cdots & \cdots & 0 & -1 \\ x_n & x_{n-1} & \cdots & \cdots & x_2 & x_1 \end{pmatrix}.$$

If $\varphi' : M_n \to Z'$ is a GL_n-invariant morphism then we define $\gamma : \mathbb{C}^n \to Z'$ by setting $\gamma = \varphi' \circ s$. One can verify that γ satisfies the required properties.

Studying the $n = 2$ case is instructive. The morphism $\varphi : M_2 \to \mathbb{C}^2$ is given by $a \mapsto (\mathrm{tr}(a), \det(a))$. A matrix $a \in M_2$ has two distinct eigenvalues if $\mathrm{tr}(a)^2 - 4\det(a) \neq 0$. Let C be the closed subset of $\mathbb{C}^2$ given by the pairs (x, y) such that $x^2 - 4y = 0$ and $p = (x, y)$ be a point of C:

(1) For $p \notin C$, the fibre $\varphi^{-1}(p)$ is the conjugate class of the diagonal matrix $\begin{pmatrix} \lambda & 0 \\ 0 & \mu \end{pmatrix}$, where (λ, μ) are the roots of the equation $t^2 - xt + y = 0$.

(2) For $p \in C$, if λ is the double root of the same equation then the fibre $\varphi^{-1}(p)$ is the union of the conjugate classes of the matrices

$$a_1 = \begin{pmatrix} \lambda & 1 \\ 0 & \lambda \end{pmatrix} \quad \text{and} \quad a_0 = \begin{pmatrix} \lambda & 0 \\ 0 & \lambda \end{pmatrix}.$$

Since the matrices $a_\epsilon = \begin{pmatrix} \lambda & \epsilon \\ 0 & \lambda \end{pmatrix}$ and a_1 are conjugate note that the closure of the orbit of a_1 contains the matrix a_0. In particular, this orbit is not closed, whereas that of a_0 is closed. Hence, there exists an open subset of the categorical quotient $\mathbb{C}^2$ of M_2 whose points correspond exactly to the orbits of the GL_2 action. Moreover, observe that $\mathbb{C}^2$ parametrizes the closed orbits of M_2.

Good quotients, geometric quotients

Definition 6.1.6 *A couple (Z, φ) consisting of an algebraic variety Z and a G-invariant morphism $\varphi : Y \to Z$ is called a* good quotient *of Y by G if it satisfies the following conditions.*

(1) *The morphism φ is affine and surjective.*

(2) *The image of a closed G-invariant sub-space of Y is closed in Z. Two closed disjoint G-invariant sub-spaces Y_1 and Y_2 of Y have disjoint images in Z.*

(3) *The structure sheaf $\mathcal{O}_Z$ of Z is the sheaf of invariant sections of the structure sheaf of Y.*

An element $f \in \Gamma(U, \varphi_*(\mathcal{O}_Y))$ is invariant under the action of G if it is invariant as a morphism $\varphi^{-1}(U) \to \mathbb{C}$. Condition (3) means that $(\varphi_*(\mathcal{O}_Y))^G = \mathcal{O}_Z$. Various more or less immediate properties of good quotients are collected together in the following proposition.

Proposition 6.1.7 *Let $\varphi : Y \to Z$ be a good quotient of Y by the action of G. Then*

(a) *the quotient (Z, φ) is categorical,*

(b) *the variety Z is equipped with the quotient topology and*

(c) *the variety Z parametrizes the closed orbit of G in Y.*

Proof The first assertion is not difficult to show but we shall not do it because the cases which will be of interest to us will be obvious. The second assertion is immediate from condition (2). For the third part note first that the orbit $\mathrm{Orb}^G(y)$ of a given point $y \in Y$ is open and everywhere dense in its closure $\overline{\mathrm{Orb}^G(y)}$. Indeed, by applying Chevalley's Theorem ([31, p.94], or [57, p.50]) to the orbit map $\sigma_g : G \to Y$, we see that there exists an open subset U of the closure of the orbit for which $U \subset \mathrm{Orb}^G(y) \subset \overline{\mathrm{Orb}^G(y)}$ and we can assume that this open subset is everywhere dense. Then we have $\mathrm{Orb}^G(y) = \bigcup_{g \in G} g.U$.

Each fibre of φ contains *at least* one closed orbit: indeed, choose an orbit O of *minimal* dimension in the set of orbits of a fibre. This is possible because φ is surjective for a good quotient. If this orbit is not closed then we consider the non-empty closed set $F = \overline{O} \setminus O$; this closed subset is G-invariant and of dimension strictly smaller that that of O. An orbit of an element of F now contradicts the minimality of the dimension of O.

We now prove that each fibre of φ contains *at most* one closed orbit. To this end, let O_1 and O_2 be two distinct orbits in the set of orbits of a fibre. Since φ is a good quotient, their images are disjoint. This is absurd. □

Definition 6.1.8 *A good quotient* (Z, φ) *of* Y *by the action of* G *is said to be* geometric *if all the orbits in* Y *under the action of* G *are closed.*

This means that if (Z, φ) is a geometric quotient then the underlying set of Z is in bijection with the set of orbits.

Lemma 6.1.9 *Let* G *be an algebraic group acting on an algebraic variety* Y *and* y *be a point of* Y*. Then the following are equivalent.*

(i) *The orbit morphism* $\sigma_y : G \to Y$ *is proper.*
(ii) *The orbit* $\mathrm{Orb}^G(y)$ *is closed in* Y *and the stabilizer* $\mathrm{Stab}^G(y)$ *is finite.*

Proof Suppose that σ_y is proper. Then the orbit $\mathrm{Orb}^G(y)$ is closed and the fibre $\sigma_y^{-1}(y) = \mathrm{Stab}^G(y)$ is finite as it is affine and complete.

Conversely, suppose that the orbit $\mathrm{Orb}^G(y)$ is closed in Y and that the stabilizer $\mathrm{Stab}^G(Y)$ is finite. Therefore, the map $G \to \mathrm{Orb}^G(y)$ is quasi-finite. There exists an open subset U of the orbit such that $\sigma_y|_{\sigma_y^{-1}(U)}\sigma_y^{-1}(U) \to \mathrm{Orb}^G(y)$ is finite. Since $\mathrm{Orb}^G(y) = \bigcup_{g \in G} g.U$ we have that σ_y is, in fact, finite and so, in particular, proper. □

Proposition 6.1.10 *Let* G *be an algebraic group operating on an algebraic variety* Y *and let* $\varphi : Y \to Z$ *be a good quotient. Let* $U \subset Y$ *be the set of the points* $y \in Y$ *such that the orbit map* $g \mapsto g.y : G \to Y$ *is proper. Then* U *is the inverse image of* φ *of an open subset* $V \subset Z$ *and*

$$\varphi|_U : U \to V$$

is a geometric quotient.

Proof Let F be the set of points $y \in Y$ such that $\dim \mathrm{Stab}^G(y) > 0$. Then F is a closed G-invariant subset. Indeed, if

$$\psi : G \times Y \longrightarrow Y \times Y$$

is the graph morphism then $\psi^{-1}(y, y) = \mathrm{Stab}^G(y) \times \{y\}$ and F is therefore closed by the theorem on the dimension of the fibres of an isomorphism (cf. Shafarevich [57]). Since $\varphi : Y \to Z$ is a good quotient, the image of F in Z is a closed set for which the complement V has an open set U as its inverse image which cannot meet F and whose points form a closed orbit: indeed, if the orbit of a point y is not closed then it would contain points in its closure whose orbits would have strictly smaller dimension. Then the image of y in Z cannot belong to V. Conversely, a point $y \in Y$ with a closed orbit and with finite stabilizer is obviously in U. Thus, by Lemma 6.1.9, the open set U is the set of points $y \in Y$ for which $g \mapsto g.y$ is proper. □

6.2 Linearly reductive groups

Let G be an affine algebraic group and let $\rho : G \to \mathrm{GL}(V)$ be a finite dimensional representation of G. Then V becomes a G-module by setting $g.v = \rho(g)v$ for $g \in G$ and $v \in V$; the globally G-invariant vector subspaces V' of V are the sub-G-modules. Recall that a representation is said to be *irreducible* if all proper G-invariant vector sub-spaces of V are trivial; in the terminology of G-modules, we say that V is a *simple* G-module. A representation ρ is said to be *totally reducible* if ρ is a direct sum of irreducible representations; we also say that V is a *semi-simple* G-module.

Definition 6.2.1 *An affine algebraic group is said to be* linearly reductive *if all (finite dimensional) G-modules V are semi-simple.*

Lemma 6.2.2 *An algebraic group G is linearly reductive if and only if for all finite dimensional representations $\rho : G \to \mathrm{GL}(V)$ every sub-G-module V' of V has a complement V'' which is also G-invariant.*

Proof Suppose otherwise. Let V be a G-module which possesses a sub-G-module V' with no complement. Choose the sub-G-module $V' \subset V$ which is maximal amongst those which have no complement. As G is linearly reductive, V is a finite direct sum of simple sub-G-modules $V = V_1 \oplus \cdots \oplus V_r$. There exists some i such that $V_i \not\subset V'$. Since V_i is simple we have that $V' \cap V_i = \{0\}$. Then $V' \oplus V_i$ is a sub-G-module for which there does not exist a complementary sub-G-module: a contradiction.

Conversely, let $V' \subset V$ be a semi-simple sub-G-module of maximal dimension. Then there exists a sub-G-module V'' such that $V = V' \oplus V''$. But then $V'' = 0$, otherwise V'' contains a non-zero simple G-module, which contradicts the maximality. □

Let Ω be the set of isomorphism classes of simple G-modules. If V is a G-module and $\omega \in \Omega$ then put

$$V_{(\omega)} = \sum_{W \subset V, W \in \omega} W.$$

If G is linearly reductive then we obtain a decomposition

$$\bigoplus_{\omega \in \Omega} V_{(\omega)} = V$$

which is called the *decomposition into isotypical factors.*

Given a direct sum decomposition $V = V^G \oplus W$, the projection V onto V^G defines a G-invariant operator $R : V \to V^G$ such that $R|_{V^G} = R$. Such an operator is called a *Reynolds operator.*

Proposition 6.2.3 *Let G be a linearly reductive group. Then, for all G-modules V, there exists one and only one Reynolds operator.*

Proof The existence follows from Lemma 6.2.2. For the uniqueness we need to show that the kernel of such a operator R is determined. But if $S \subset V$ is a non-trivial simple sub-module then the restriction $R|_S$ is either injective or zero since S is simple; the first case is obviously excluded. It follows that

$$\ker R = \bigoplus_{\omega \neq 0} V_{(\omega)},$$

where 0 denotes the isomorphism class of the trivial simple G-module. From this we have the uniqueness. □

Proposition 6.2.4 *The Reynold's operator commutes with G-module maps, i.e. if $f : V \to W$ is a G-module map then the diagram*

$$\begin{array}{ccc} V & \xrightarrow{f} & W \\ {\scriptstyle R}\downarrow & & \downarrow{\scriptstyle R} \\ V^G & \xrightarrow{f} & W^G \end{array}$$

commutes.

Proof From the description of the kernel of the Reynold's operator given above, we have $f(\ker R) \subset \ker R$. It follows that $R \circ f$ takes the value 0 on $\ker R$ and coincides with f on V^G. Hence, $f \circ R = R \circ f$. □

Proposition 6.2.5 *Let G be an algebraic group and suppose that there is a Reynold's operator for each G-module V which commutes with the G-module maps. Then G is linearly reductive.*

Proof Let V and W be two G-modules. Then $\mathrm{Hom}(V, W)$ is a G-module by the action $g.f : v \mapsto g.f(g^{-1}.v)$. Note that f is a G-module map if and only if f is G-invariant. By Lemma 6.2.2 we must show that whenever we are given a sub-G-module $V' \subset V$ there exists a complementary sub-G-module V'' to V'. Let $\pi : V \to V'$ be the projection map of V to V'. By the hypothesis, there exists a Reynold's operator on the G-module $\mathrm{Hom}(V, V')$ and $R(\pi)$ is then a G-module map. From the commutation property it follows that $R(\pi)|_{V'} = \mathrm{id}_{V'}$ and so we have found the sub-module $V'' = \ker R(\pi)$. □

Proposition 6.2.6 *Let G be a linearly reductive group. The functor which associates the vector space V^G of invariants to a G-module V is exact.*

Proof In full generality, the functor is left exact. We need to verify that if $f : V \to W$ is a surjective G-module map then the induced map $V^G \to W^G$ is also surjective if the group G is linearly reductive. Let $y \in W^G$. There exists $x \in V$ such that $f(x) = y$. Then by Proposition 6.2.4, we have $f(R(x)) = y$ from which the surjectivity follows. □

Examples

(1) Let G be a finite group, $\#G$ the number of its elements and V a G-module. Then the operator R defined by

$$R(v) = \frac{1}{\#G} \sum_{g \in G} g.v$$

is the Reynold's operator. Therefore all finite groups are linearly reductive by Proposition 6.2.5. This is Maschke's Theorem.
(2) Let K be compact topological group. Then K has a unique measure called the *Haar measure* which satisfies the following conditions.

(I) right and left invariance: if f is measurable then

$$\int_K f(k)\, dk = \int_K f(g^{-1}k)\, dk = \int_K f(kg)\, dk$$

for all $g \in G$.
(N) Normalization:

$$\int_K dk = 1.$$

If V is a (finite dimensional) K-module then the operator R defined by $R(v) = \int_K k.v\, dk$ is the Reynold's operator. Therefore, every K-module V is semi-simple. We can deduce the linear reductivity of the classical groups ("Weyl's unitary trick"):

Lemma 6.2.7 *Let G be an algebraic group. Suppose that G contains a compact sub-group K (for the usual topology) which is Zariski dense in G. Then G is linearly reductive.*

Proof Let V be a G-module and $V' \subset V$ a sub-G-module. The G-module V is a K-module and V' is preserved by K. Then the sub-module V' admits a complement V'' which is also preserved by K. Let N be the sub-group of elements of G which leave V'' globally invariant. The sub-group N is (Zariski) closed and contains K. Then, by the hypothesis, we have $N = G$. Hence, V'' is a G-invariant complement, which proves the lemma. □

For example, for $\mathrm{GL}_n(\mathbb{C})$ we reason in the following way. Let U_n be the unitary group. On the level of Lie algebras we have

$$\mathrm{Lie}\, U_n \otimes_{\mathbb{R}} \mathbb{C} = \mathrm{Lie}\, \mathrm{GL}_n(\mathbb{C}).$$

Let H be the Zariski closure of U_n in $\mathrm{GL}_n(\mathbb{C})$. This is an algebraic sub-group and we have $\mathrm{Lie}(H)$ $=\mathrm{Lie}(\mathrm{GL}_n)$ and, since GL_n is connected, we have $H = \mathrm{GL}_n$. The unitary group is then a sub-group of $\mathrm{GL}_n(\mathbb{C})$ which is compact and everywhere dense in the Zariski topology, and so the hypothesis of the lemma is satisfied.
(3) Let $(\mathbb{C}, +)$ be the additive group:

$$(\mathbb{C}, +) = \left\{ \begin{pmatrix} 1 & \star \\ 0 & 1 \end{pmatrix} \subset \mathrm{GL}_2(\mathbb{C}) \right\}.$$

Exercise: The group $(\mathbb{C}, +)$ is not linearly reductive.

In general, we say that an algebraic group is *unipotent* if it is isomorphic to a closed sub-group of the sub-group of elements of $\mathrm{GL}_n(\mathbb{C})$ of matrices of the form

$$\begin{pmatrix} 1 & \star & \star \\ 0 & \ddots & \star \\ 0 & 0 & 1 \end{pmatrix}.$$

Exercise: A non-trivial unipotent group is never linearly reductive.

6.3 The affine case

The variety Spec A

For any algebraic variety S we let $\mathcal{O}(S)$ denote the coordinate algebra of S, i.e. the algebra of morphisms $S \to \mathbb{C}$. Recall that if we are given an algebra A of finite type over $\mathbb{C}$ then the contravariant functor defined on the category of affine algebraic varieties which takes an algebra A to the set $\mathrm{Hom}_{\mathrm{alg}}(A, \mathcal{O}(S))$ of algebra homomorphisms of A to $\mathcal{O}(S)$ is representable by an algebraic variety, denoted by $\operatorname{Spec} A$. This means that there is a functorial isomorphism

$$\mathrm{Mor}(S, \operatorname{Spec} A) \simeq \mathrm{Hom}_{\mathrm{alg}}(A, \mathcal{O}(S)).$$

The closed points of this variety can be identified with the maximal ideal of A and the canonical map $A \to \mathcal{O}(\operatorname{Spec} A)$ is an isomorphism. If $(x_1, \ldots, x_n)$ is a system of generators of A related by the polynomials $P_1(x_1, \ldots, x_n) = 0, \ldots, P_k(x_1, \ldots, x_n) = 0$ then the variety $\operatorname{Spec} A$ is isomorphic to the sub-variety of $\mathbb{C}^n$ defined by the ideal generated by the polynomials $(P_1, \ldots, P_k)$.

Passing to the quotient

Theorem 6.3.1 (Hilbert-Mumford) *Let G be a linearly reductive algebraic group acting on the affine algebraic variety $Y = \operatorname{Spec} A$. Then the $\mathbb{C}$-algebra A^G is of finite type. Let $Z = \operatorname{Spec} A^G$ and $\varphi : Y \to Z$ be the morphism induced by the inclusion $A^G \to A$. Then the pair (Z, φ) is a good quotient.*

For the proof we need some preliminaries. Suppose that G acts on the affine variety Y. Then G also acts on $\mathcal{O}(Y)$ by

$$f \mapsto g^*(f) = f \circ g^{-1}$$

and $\mathcal{O}(Y)$ then becomes a G-module, generally of infinite dimension. This leads to the following definition.

Definition 6.3.2 *Let V be a G-module, perhaps infinite dimensional. We say that V is* locally finite *if for every finite dimensional vector sub-space $V' \subset V$, the G-module it generates, denoted by $G.V'$, is finite dimensional.*

Lemma 6.3.3 *Suppose that G acts on the affine variety Y. Then the G-module $\mathcal{O}(Y)$ is locally finite.*

Proof The operation $\sigma : G \times Y \to Y$ induces a $\mathbb{C}$-algebra map

$$\sigma^* : \mathcal{O}(Y) \to \mathcal{O}(G) \otimes \mathcal{O}(Y).$$

Let $f \in \mathcal{O}(Y)$. We show that the orbit of f is contained in a finite dimensional vector sub-space. Write

$$\sigma^*(f) = \sum_{i=1}^{n} \varphi_i \otimes f_i$$

with $\varphi_i \in \mathcal{O}(G)$ and $f_i \in \mathcal{O}(Y)$. We show that the orbit of f is contained in the vector sub-space generated by $(f_1, \dots, f_n)$. The map σ is G-equivariant when we equip the product $G \times Y$ with the action given by left multiplication on G and the trivial action on the Y factor; by transposition, we see that the map σ^* is also G-equivariant. Then we have

$$\sigma^*(g^*(f)) = \sum_{i=1}^{n} g^*(\varphi_i) \otimes f_i.$$

Composing σ with the inclusion $Y \hookrightarrow G \times Y$ associated to the identity element of G, we get the identity on Y. It follows that $g^*(f) = \sum_{i=1}^{n} \varphi_i(g^{-1}).f_i$, which proves the result. □

In virtue of this lemma, we can apply the preceding results on linearly reductive groups to the G-module $\mathcal{O}(Y)$. In particular, there exists a G-linear projection $R : \mathcal{O}(Y) \to \mathcal{O}(Y)^G$ which we still call the Reynolds operator.

Lemma 6.3.4 *Let G be a linearly reductive algebraic group operating on an affine variety Y and let R be the Reynolds operator. Then R is $\mathcal{O}(Y)^G$-linear.*

Proof Let $f \in \mathcal{O}(Y)^G$. The multiplication map given by f defines a G-module map $\mathcal{O}(Y) \to \mathcal{O}(Y)$. Then it follows from Proposition 6.2.4, extended to the case of G-modules which are locally of finite dimension, that we have $R(fh) = fR(h)$ for all $h \in \mathcal{O}(Y)$. □

The proof of the theorem will follow from the next fact.

Lemma 6.3.5 *Let G be a linearly reductive algebraic group acting on the affine variety Y. If J is an ideal of $\mathcal{O}(Y)^G$ then we have $J\mathcal{O}(Y) \cap \mathcal{O}(Y)^G = J$.*

Proof It is clear that the ideal J is contained in the left hand side. For the other way around, let $f \in J\mathcal{O}(Y)$ be invariant under the action of G. Then we have $f = \sum_i \alpha_i g_i$ with $\alpha_i \in J$ and $g_i \in \mathcal{O}(Y)$. If we apply the Reynolds operator we get, taking Lemma 6.3.4 into account, $f = R(f) = \sum_i \alpha_i R(g_i)$ and so f is in J. □

Proof of Theorem 6.3.1 We show that A^G is a $\mathbb{C}$-algebra of finite type. First we shall study the particular case of a G-module V. The polynomial algebra $A = \mathcal{O}(V)$ is graded: we have

$$A = \bigoplus_{i \geq 0} A_i,$$

where A_i is the vector sub-space of homogeneous polynomials of degree i. This grading is compatible with the G-module structure. It follows that the Reynolds operator is compatible with this grading and the induced grading on the $\mathbb{C}$-algebra $B = \mathcal{O}(V)^G$ (so $B_i = A_i \cap B$). By Hilbert's Theorem, the ring A is Noetherian and the homogeneous ideal AB_+, with $B_+ = \bigoplus_{i>0} B_i$, is of finite type; i.e. AB_+ is generated by a finite number of elements $z_1, \ldots, z_r$ which we can suppose to be homogeneous and belong to B_+. Let ν_i be the degree of z_i. We now show by induction on the degree that each of the elements of B is a polynomial in z_i with coefficients in $\mathbb{C}$. Indeed, this is true in degree 0 because $B_0 = \mathbb{C}$. If $b \in B_i$, with $i > 0$, then $b = \sum_{i=1}^r a_i z_i$ with $\deg(a_i) < i$. So we have $b = R(b) = \sum_{i=1}^r R(a_i z_i) = \sum_{i=1}^r R(a_i) z_i$ with $R(a_i) \in B$ of degree $d_i = \deg(a_i) < i$.

By induction, we see that

$$b_i = R(a_i) = \sum_{\ell_1 \nu_1 + \cdots + \ell_r \nu_r = d_i} \alpha_{\ell_1, \ldots, \ell_r} z_1^{\ell_1} \cdots z_r^{\ell_r}$$

with $\alpha_{\ell_1, \ldots, \ell_r} \in \mathbb{C}$. Consequently, b is a polynomial in z_i with complex coefficients.

The general case reduces to the particular case. Indeed, we show that if the algebraic group G acts on the affine variety Y then there exist a G-module V and a G-equivariant embedding of Y into V. By Lemma 6.3.3, there exists a finite dimensional sub-G-module W of $\mathcal{O}(Y)$ which generates $\mathcal{O}(Y)$ as an algebra. The inclusion $W \hookrightarrow \mathcal{O}(Y)$ induces a surjection of algebras $\mathrm{Sym}(W) \to \mathcal{O}(Y)$, compatible with the G action. The variety $\mathrm{Spec}(\mathrm{Sym}(W))$ is given a G action and can be identified with the dual G-module $V = W^*$; the surjection above induces a G-equivariant embedding $Y \hookrightarrow V$ which is what we wanted.

The $\mathbb{C}$-algebra $\mathcal{O}(V)^G$ is of finite type by the particular case and the morphism $\mathcal{O}(V) \to \mathcal{O}(Y)$ is a surjection of locally finite G-modules; after Proposition 6.2.6, we obtain a surjection $\mathcal{O}(V)^G \to \mathcal{O}(Y)^G$ and hence, the $\mathbb{C}$-algebra $\mathcal{O}(Y)^G$ is also of finite type.

Let $Z = \operatorname{Spec} A^G$ and let $\varphi : Y \to Z$ be the map defined by the inclusion $A^G \hookrightarrow A$. We verify that this is a good quotient.

(1) We show that φ surjects. By Lemma 6.3.5, we have

$$J\mathcal{O}(Y) \cap \mathcal{O}(Y)^G = J$$

for all ideals J of $\mathcal{O}(Y)^G$. This implies that we have $\overline{\varphi(\varphi^{-1}(K))} = K$ for each closed subset K of Z. If we apply this result to the closed subsets given by points of Z then we obtain the surjection.

(2) We show that the image of a closed invariant subset F of Y is a closed subset of Z. If I is the ideal associated to F then I is G-invariant and the quotient A/I is the algebra of finite type associated to F considered as an affine variety. The group G acts on F and, by what we have proved, the canonical map

$$F = \operatorname{Spec}(A/I) \longrightarrow \operatorname{Spec}(A/I)^G$$

surjects. But, by Proposition 6.2.6, the map $A^G/I^G \to (A/I)^G$ is an isomorphism which shows that $\operatorname{Spec}(A/I)^G$ is the closed subset of Z given by the ideal I^G. Moreover, the above map is none other than the restriction of φ to F. Hence, $\varphi(F)$ is closed.

(3) Let F_1 and F_2 be two disjoint closed G-invariant subsets of Y. Then there exists a morphism $f \in \mathcal{O}(Y)$ which takes the value 0 on F_1 and 1 on F_2. Consider the morphism $g = R(f)$. By Proposition 6.2.4, this morphism still takes the value 0 on F_1 and 1 on F_2. Thus, g defines a morphism of $\mathcal{O}(Z)$ which separates the closed subsets $\varphi(F_1)$ and $\varphi(F_2)$.

It remains to show, as promised, that it is a categorical quotient. If we have chosen an embedding $Z \hookrightarrow \mathbb{C}^p$ then note that the morphism φ is given by a G-invariant morphism in $\mathcal{O}(Y)^G$; it follows that it is G-equivariant. Let (Z', φ') be another closed pair consisting of an algebraic variety and an equivariant morphism $\varphi' : Y \to Z'$. Suppose first of all that Z' is affine. Then φ' induces a map of algebras $\mathcal{O}(Z') \to \mathcal{O}(Y)$ which factors through $\mathcal{O}(Z) = \mathcal{O}(Y)^G$. This gives us a map $\gamma : Z \to Z'$ such that $\gamma \circ \varphi = \varphi'$. Suppose now that Z is arbitrary. Cover Z' with open affine subsets V' and note that the open sets $U = \varphi'^{-1}(V')$ are then G-equivariant. Using the separation property of closed G-invariant subsets, we can cover Z with principal affine open sets D_f, with $f \in A^G$, such that each D_f is contained in such an open subset U. Then, using the algebra of fractions $(A_f)^G \simeq (A^G)_f$, the induced map $\varphi' : D_f \to V'$ factorizes as in the following diagram:

$$\begin{array}{ccc} D_f = \operatorname{Spec} A_f & \xrightarrow{\varphi} & \operatorname{Spec}(A^G)_f \\ & \searrow^{\varphi'} & \downarrow{\scriptstyle \gamma_f} \\ & & V' \subset Z' \end{array}$$

It remains to note that the maps γ_f patch together to give a map $\gamma : Z \to Z'$. This completes the proof. □

6.4 The projective case

The variety $\operatorname{Proj} A$

Consider in this section a graded $\mathbb{C}$-algebra of finite type $A = \bigoplus_{i \geq 0} A_i$ satisfying the following condition:

$$(*) \qquad A_0 = \mathbb{C}.$$

Such an algebra is equipped with a $\mathbb{C}^*$ action by setting $t.f = t^{-i} f$ for $f \in A_i$ and $t \in \mathbb{C}^*$. It follows that the variety $\operatorname{Spec} A$ is also equipped with a $\mathbb{C}^*$ action; the only fixed point is 0 corresponding to the maximal ideal $A_+ = \bigoplus_{i>0} A_i$ of A. The open complement $\operatorname{Spec}^0(A)$ of $\{0\}$ is not an affine variety in general. However, there exists a geometric quotient of $\operatorname{Spec}^0(A)$ by $\mathbb{C}^*$.

Proposition 6.4.1 *(1) The $\mathbb{C}^*$ action on $\operatorname{Spec}^0(A)$ is proper.*

(2) The orbit space $\operatorname{Spec}^0(A)/\mathbb{C}^*$, equipped with the quotient topology and the sheaf of invariants, is an algebraic variety.

(3) This algebraic variety is projective.

Proof (1) let $\zeta_1, \ldots, \zeta_n$ be the homogeneous generators of A with $\zeta_i \in A_{m_i}$. These generators define a closed embedding $\operatorname{Spec} A \hookrightarrow \mathbb{C}^n$ which is $\mathbb{C}^*$-equivariant if we give $\mathbb{C}^n$ the $\mathbb{C}^*$ action defined by

$$t.(z_1, \ldots, z_n) = (t^{m_1} z_1, \ldots, t^{m_n} z_n).$$

The action of $\mathbb{C}^*$ on $\mathbb{C}^n \setminus \{0\}$ is proper. It follows that the action of $\mathbb{C}^*$ on $\operatorname{Spec}^0(A)$ is proper.

(2) We have open subsets D_i of $\operatorname{Spec} A$ defined by $\zeta_i \neq 0$. We also have $D_i = \operatorname{Spec} A \cap \{z_i \neq 0\}$. These open subsets D_i are affine and

$\mathbb{C}^*$-invariant, so there exists a good quotient of D_i by $\mathbb{C}^*$. Since the action is proper, these quotients are geometric. The open sets D_i cover $\mathrm{Spec}^0(A)$, which proves that the topological space $\mathrm{Spec}^0(A)/\mathbb{C}^*$ with structure sheaf given by the sheaf of invariant sections is locally isomorphic to an affine variety. The separation condition which we require of algebraic varieties is satisfied because the $\mathbb{C}^*$ action is proper and the finiteness condition is trivial, so we have an algebraic variety.

(3) Let $m = \mathrm{lcm}(m_i)$ and $\nu_i = m/m_i$. The map $\mathrm{Spec}\, A \to \mathbb{C}^n$ defined by $(z_1, \ldots, z_n) \mapsto (z_1^{\nu_1}, \ldots, z_n^{\nu_n})$ induces a finite morphism

$$\mathrm{Spec}^0(A)/\mathbb{C}^* \longrightarrow \mathbf{P}_{n-1}(\mathbb{C}).$$

It follows that $\mathrm{Spec}^0(A)/\mathbb{C}^*$ is a projective variety. □

Definition 6.4.2 *Given a graded $\mathbb{C}$-algebra A of finite type such that $A_0 = \mathbb{C}$, the projective variety* $\mathrm{Spec}^0(A)/\mathbb{C}^*$ *is denoted* $\mathrm{Proj}\, A$.

Let A and B be two graded algebras of finite type satisfying condition $(*)$ and let $\varphi : A \to B$ be a degree preserving homomorphism. This defines a $\mathbb{C}^*$-equivariant map of affine varieties $\varphi^* : \mathrm{Spec}\, B \to \mathrm{Spec}\, A$. The fibre ${\varphi^*}^{-1}(0)$ over 0 is given by the homogeneous ideal $I = \varphi(A_+)B$ of B; it is invariant under $\mathbb{C}$ and gives rise to a closed subset $V(I)$ in $\mathrm{Proj}\, B$ which is not, in general, empty. The map φ^* induces a morphism of algebraic varieties

$$\phi : \mathrm{Proj}\, B \setminus V(I) \longrightarrow \mathrm{Proj}\, A.$$

The closed set $V(I)$ will be called the *centre* of the map φ.

The existence of a good quotient

Let G be a linearly reductive algebraic group acting linearly on $\mathbb{C}^{n+1}$. This action commutes with that of $\mathbb{C}^*$ so we see that G also acts on $\mathbf{P}_n$. Let $Y \subset \mathbf{P}_n$ be a closed G-invariant sub-variety given by a G-invariant homogeneous ideal $I \subset \mathcal{O}(\mathbb{C}^{n+1})$. The $\mathbb{C}$-algebra $A = \mathcal{O}(\mathbb{C}^{n+1})/I$ is a graded algebra of finite type satisfying $(*)$ and Y can be identified with $\mathrm{Proj}\, A$. By Theorem 6.3.1, the algebra A^G is also an algebra of finite type; we give it the induced grading.

Definition 6.4.3 *A point $a \in Y$ is said to be* semi-stable *under the action of G if there exists a homogeneous G-invariant polynomial with strictly positive degree which does not vanish at a. Otherwise, such a point is said to be* unstable *for the G action.*

The closed subset of unstable points is just the centre $V(A^G_+)$ of the inclusion $A^G \hookrightarrow A$. Let Y^{ss} be the open subset of semi-stable points of Y and $Z = \operatorname{Proj} A^G$. Then we have a morphism of algebraic varieties

$$Y^{ss} \longrightarrow \operatorname{Proj} A^G.$$

Theorem 6.4.4 *The morphism $\phi : Y^{ss} \to Z = \operatorname{Proj} A^G$ induced by the inclusion $A^G \subset A$ is a good quotient of the open set Y^{ss} of semi-stable points of Y by the action of G.*

Proof The question is local on Z so we cover Z with affine open subsets V_i such that $U_i = \phi^{-1}(V_i)$ are affine and such that $\varphi : U_i \to V_i$ is a good quotient. We have seen that if $\zeta_1, \ldots, \zeta_k$ generate A^G then the variety Z can be covered by the open affine subsets $V_i = \{\zeta_i \neq 0\}/\mathbb{C}^*$ and the inverse image of this open subset is the affine subset U_i of Y defined in the same way. The coordinate algebra of U_i is the algebra $A_{(\zeta_i)}$ of fractions of degree 0 with denominator a power of ζ_i. Similarly, the coordinate algebra of V_i is the algebra $A^G_{(\zeta_i)}$ of degree zero fractions of elements of A^G. This algebra is certainly isomorphic to the algebra of invariants ${G_{(\zeta_i)}}^G$ and the morphism $\phi|_{U_i} : U_i \to V_i$ is obviously induced by the inclusion ${A_{(\zeta_i)}}^G \hookrightarrow A_{(\zeta_i)}$; by Theorem 6.3.1, it is a good quotient. □

Stable and unstable points

We let $\hat{Y}$ denote the cone on Y. So we have

$$\hat{Y} = \operatorname{Spec} A \qquad \text{and} \qquad \hat{Y} \setminus \{0\} = \operatorname{Spec}^0 A.$$

Lemma 6.4.5 *A point $a \in Y$ is unstable under the G action if and only if it is the image of a point $a' \in \hat{Y} \setminus \{0\}$ such that $\overline{\operatorname{Orb}^G(a')}$ contains 0.*

Proof Let $a' \in \hat{Y} \setminus \{0\}$ correspond to $a \in Y$. All G-invariant functions are constant on $\overline{\operatorname{Orb}^G(a')}$. If 0 is contained in $\overline{\operatorname{Orb}^G(a')}$ then such a function, if it is homogeneous of strictly positive degree, must vanish at 0 and so also at a'. So the point a is unstable.

If 0 is not contained in $\overline{\operatorname{Orb}^G(a')}$ then we can separate the closed G-invariant sets $\{0\}$ and $\overline{\operatorname{Orb}^G(a')}$ by a G-invariant polynomial $f \in A^G$ which takes the values 0 at 0 and 1 on $\overline{\operatorname{Orb}^G(a')}$ (cf. proof of Theorem 6.3.1). But then we can find a G-invariant homogeneous polynomial of degree $k > 0$ which does not vanish on $\overline{\mathrm{Orb}^G(a')}$ and, in particular, at a': so a is semi-stable. $\square$

Definition 6.4.6 *A point $a \in Y$ is said to be* stable[1] *under the action of G if it is semi-stable and if the orbit map $g \mapsto g.a : G \to Y^{ss}$ is proper.*

From Proposition 6.1.10 we deduce

Proposition 6.4.7 *The set Y^s of stable points of Y is the inverse image under φ of the open set Z^s of Z and the induced map $\varphi : Y^s \to Z^s$ is a geometric quotient.*

We can describe the stable points in terms of the cone on Y:

Proposition 6.4.8 *Let a be a point of Y which is the image of a point $a' \in \hat{Y} \setminus \{0\}$. Then $a \in Y$ is stable if and only if the orbit map $g \mapsto g.a' : G \to \hat{Y}$ is proper.*

Proof Let $\hat{Y}^{ss}$ denote the inverse image of Y^{ss} in the cone. The commutative diagram

$$\begin{array}{ccc} G & \xrightarrow{\sigma_{a'}} & \hat{Y}^{ss} \\ & \searrow^{\sigma_a} & \downarrow \\ & & Y^{ss} \end{array}$$

shows that if $g \mapsto g.a : G \to Y^{ss}$ is proper then so is $g \mapsto g.a' : G \to \hat{Y}^{ss}$. So the stabilizer of a' is finite and the closure of the orbit $\mathrm{Orb}^G(a')$ in $\hat{Y}$ does not contain 0 because a is semi-stable by definition. It follows that the orbit $\mathrm{Orb}^G(a')$ is closed in $\hat{Y}$. From Lemma 6.1.9 it follows that the orbit map $g \mapsto g.a' : G \to \hat{Y}$ is proper.

Conversely, if $g \mapsto g.a' : G \to \hat{Y}$ is proper then the point a is obviously semi-stable, the stabilizer $\mathrm{Stab}^G a'$ is finite and we have an exact sequence of groups

$$1 \longrightarrow \mathrm{Stab}^G a' \longrightarrow \mathrm{Stab}^G a \longrightarrow \mathbb{C}^*.$$

But the image of $\mathrm{Stab}^G a$ in $\mathbb{C}^*$ is finite, otherwise 0 would be contained in the closure of the orbit, therefore contained in the orbit, and so $a' = 0$.

[1] Mumford called this *properly stable*

Hence, $\mathrm{Stab}^G a$ is finite. It remains to see that the orbit $\mathrm{Orb}^G a$ is closed in Y^{ss}. Let $\hat{Z} = \mathrm{Spec}\, A^G$. Consider the commutative diagram

$$\begin{array}{ccc} \hat{Y}^{ss} & \longrightarrow & \hat{Z} \setminus \{0\} \\ \downarrow & & \downarrow \\ Y^{ss} & \xrightarrow{\varphi} & Z \end{array}$$

The orbit $\mathrm{Orb}^G a'$ of a' defines a point $b' \in \hat{Z} \setminus \{0\}$ and we denote its class in Z by b. The inverse image of b in Y^{ss} is exactly the orbit $\mathrm{Orb}^G b$; indeed, it is made up of the images of points a'' of $\mathrm{Orb}^{\mathbb{C}^* \times G} a'$. Hence, the orbit $\mathrm{Orb}^G a$ is closed in Y^{ss}. So the point a is stable. □

This leads to the following definition:

Definition 6.4.9 *Let V be a G-module and let v be a point of V. We say that v is*

- semi-stable *if* $0 \notin \overline{\mathrm{Orb}^G(v)}$,
- stable *if the orbit* $\mathrm{Orb}^G(v)$ *is closed and the stabilizer* $\mathrm{Stab}^G(v)$ *is finite.*

6.5 The Hilbert-Mumford criterion

In the case where the group G is finite then all of the points are stable and this gives us a geometric quotient which is a projective variety. In the general case, to be able to compute effectively the semi-stable or stable points of the action of a linearly reductive algebraic group G on a projective variety $Y \subset \mathbf{P}_n$ which extends to $V = \mathbb{C}^{n+1}$, we need to have at our disposal a highly useful numerical criterion called the Hilbert-Mumford criterion.

Definition 6.5.1 *A non-trivial morphism of algebraic groups from the multiplicative group $G_m = \mathbb{C}^*$ to an algebraic group G is called a* one parameter sub-group *of G.*

Let Y be a quasi-projective algebraic variety on which the algebraic group G acts. Each one parameter sub-group λ of G induces an action of $G_m = \mathbb{C}^*$ on Y by composition. The group $\mathbb{C}^*$ can be considered as an open subset of the projective line $\mathbf{P}_1$ by identifying $x \in \mathbb{C}$ with $[x : 1]$. We let ∞ denote the point $[1 : 0]$. If $\overline{Y}$ is a completion of Y and y a point of Y then the map $\mathbb{C}^* \to \overline{Y} : t \mapsto \lambda(t).y$ extends in a unique way to the projective line $\mathbf{P}_1$ and this gives meaning to

$$y_0 = \lim_{t \to 0} \lambda(t).y \qquad \text{and} \qquad y_\infty = \lim_{t \to \infty} \lambda(t).y.$$

These points are fixed under the action of the given one parameter sub-group.

Lemma 6.5.2 *Let $y \in Y$. The map $\lambda_y : t \mapsto \lambda(t)y : \mathbb{C}^* \to Y$ is proper if and only if y_0 and y_∞ are not contained in Y.*

Proof Suppose, for example, that y_0 is contained in Y. Then the orbit $\mathrm{Orb}^{\mathbb{C}^*} y$ is only closed if y is fixed by the action of λ. By Lemma 6.1.9, the map λ_y is not proper. On the other hand, if y_0 and y_∞ are not contained in Y then the inverse image of Y under the morphism $\bar{\lambda}_y : \mathbf{P}_1 \to \overline{Y}$ which extends λ_y is exactly $\mathbb{C}^*$. Since $\bar{\lambda}_y$ is proper we have that λ_y is also proper. □

We state the following result:

Theorem 6.5.3 (Hilbert-Mumford) *Let G be a linearly reductive group and V a G-module. A point $v \in V$ is semi-stable (respectively stable) under the action of G if and only if for* all *one parameter sub-groups it is semi-stable (respectively stable) under the induced $\mathbb{C}^*$ action.*

Lemma 6.5.4 *Let V be a G-module and $\lambda : \mathbb{C}^* \to G$ be a one parameter sub-group. For each integer $m \in \mathbb{Z}$, let $V_m = \bigcap_t \ker(\lambda(t) - t^m \mathrm{id}_V)$. Then*

$$\bigoplus_{m \in \mathbb{Z}} V_m = V.$$

Proof We have seen that any action of $\mathbb{C}^*$ on a projective variety admits at least one fixed point. It follows that there exists a line $V' \subset V$ which is invariant under the $\mathbb{C}^*$ action. Since $\mathbb{C}^*$ is a reductive group, we can find a complement to V' which is invariant under $\mathbb{C}^*$ and then we induct on the dimension of V. □

For $v \in V$, we write $v = \sum_{m \in \mathbb{Z}} v_m$ and set

$$\mu(\lambda, v) = -\min\{m \mid v_m \neq 0\}.$$

Suppose now that G acts linearly on $V = \mathbb{C}^{n+1}$. Let $Y \subset \mathbf{P}_n$ be a closed G-invariant sub-variety and let $y \in Y$. We put $\mu(\lambda, y) = \mu(\lambda, y')$, where $y' \in \hat{Y} \setminus \{0\}$ is a point of the cone on Y projecting to y.

Note the following:

(i) $\mu(\lambda, y) > 0$ if and only if $\lim_{t\to 0}\lambda(t).(y')$ does not exist in $\hat{Y}$.
(ii) $\mu(\lambda, y) < 0$ if and only if $\lim_{t\to 0}\lambda(t).(y') = 0$.

Theorem 6.5.5 (The Hilbert-Mumford criterion) *Suppose that G acts linearly on $\mathbb{C}^{n+1}$. Let $Y \subset \mathbf{P}_n$ be a closed G-invariant sub-variety. A point $y \in Y$ is semi-stable (respectively stable) if and only if for each one parameter sub-group λ of G, we have $\mu(\lambda, y) \geq 0$ (respectively $\mu(\lambda, y) > 0$).*

Proof We can immediately reduce to the case of $\mathbf{P}_n$ and hence to the case of a G-module V. By Theorem 6.5.3, a vector v is semi-stable under the action of G if and only if v is semi-stable under the induced action of $\mathbb{C}^*$ for all one parameter sub-groups of G. This is equivalent to $\mu(\lambda, v) \geq 0$ for any one parameter sub-groups λ. Also from Theorem 6.5.3, a vector v is G-stable if and only if it is $\mathbb{C}^*$-stable for each one parameter sub-group λ of G, which is equivalent to $\mu(\lambda, v) > 0$ for each one parameter sub-group λ. □

6.6 A computation of the semi-stable points

Let H and V be two $\mathbb{C}$-vector spaces of dimension n and q, respectively. Let $\mathbf{Grass}(m, H \otimes V)$ denote the Grassmannian of vector sub-spaces of dimension m in $H \otimes V$. The linearly reductive algebraic group $\mathrm{SL}(H)$ acts on $H \otimes V$ by $g.(h \otimes v) = g.h \otimes v$. It follows that the group $\mathrm{SL}(H)$ acts on $\mathbf{Grass}(m, H \otimes V)$ by setting $g.K = (g \otimes \mathrm{id})(K)$. We also get a linear action of $\mathrm{SL}(H)$ on $\Lambda^m(H \otimes V)$ by $g.(x_1 \wedge \cdots \wedge x_m) = g.x_1 \wedge \cdots \wedge g.x_m$ and so an action on the projective space $\mathbf{P}(\Lambda^m(H \otimes V))$. The Plücker embedding

$$\mathbf{Grass}(m, H \otimes V) \hookrightarrow \mathbf{P}(\Lambda^m(H \otimes V))$$

is then $\mathrm{SL}(H)$-equivariant. We aim to determine the semi-stable and stable points of the $\mathrm{SL}(H)$ action on $\mathbf{Grass}(m, H \otimes V)$ considered as a closed invariant sub-variety of this projective space under the Plücker embedding.

Proposition 6.6.1 *Let $K \in \mathbf{Grass}(m, H \otimes V)$. The following are equivalent:*

(i) the point K is semi-stable (respectively stable) under $SL(H)$;

(ii) *for each proper non-zero vector sub-space $H' \subset H$ we have, setting $K' = (H' \otimes V) \cap K$,*

$$\frac{\dim K'}{\dim H'} \leq \frac{\dim K}{\dim H} \qquad (\textit{respectively} <.)$$

Proof We must compute $\mu(\lambda, K)$, for all one parameter sub-groups λ of $\mathrm{SL}(H)$. Let λ be a one parameter of $\mathrm{SL}(H)$. We can write

$$H = \bigoplus_{i=1}^{s} H_i$$

where H_i is a direct sum of examples of the irreducible representation of $\mathbb{C}^*$ of weight r_i, given by $t \mapsto t^{r_i}$. We have $\sum_{i=1}^{s} r_i \dim H_i = 0$, and we can suppose that $r_i > \ldots > r_s$. Consider the increasing filtration of H defined by $F_i = \oplus_{j \leq i} H_j$ and the induced filtration $K_i = (F_i \otimes V) \cap K$.

Lemma 6.6.2 *We have* $\mu(\lambda, K) = \sum\limits_{i=1}^{s} r_i \dim(K_i/K_{i-1})$.

Proof Consider the decomposition of the vector space

$$\wedge^m(H \otimes V) = \bigoplus_{j_1+\ldots+j_s=m}^{j_1,\ldots,j_s} \wedge^{j_1}(H_1 \otimes V) \otimes \ldots \otimes \wedge^{j_s}(H_s \otimes V).$$

The associated graded vector space $\mathrm{gr.}(K)$ is a graded vector sub-space of the graded vector space $H \otimes V$. Let $\nu_i = \dim \mathrm{gr}_i(K)$. Then we have

$$\wedge^m \mathrm{gr.}(K) \subset \wedge^{\nu_1}(H_1 \otimes V) \otimes \ldots \otimes \wedge^{\nu_s}(H_s \otimes V)$$

and then this line in $\wedge^m(H \otimes V)$ is globally $\mathbb{C}^*$-invariant with weight $\sum_{i=1}^{s} r_i \nu_i$. Let $\bar{K} = \lim_{t \to 0} \lambda(t) K$. Then $\bar{K}$ is $\mathbb{C}^*$-invariant, and we have a $\mathbb{C}^*$-isomorphism $\bar{K} \xrightarrow{\sim} \mathrm{gr.}(\bar{K})$. That implies that $\mu(\lambda, K) = \mu(\lambda, \bar{K}) = -\sum_{i=1}^{s} r_i v_i$. □

End of the proof of proposition 6.6.1 By the Abel transformation, we have

$$-\mu(\lambda, K) = \sum_{i=1}^{s-1} \dim K_i (r_i - r_{i+1}) + r_s m.$$

For $H' \in H$, with $H' \neq \{0\}$ and $H' \neq H$ we put $K' = (H' \otimes V) \cap K$. Suppose that

$$\frac{\dim K'}{\dim H'} \leq \frac{\dim K}{\dim H} \text{ (resp. <).}$$

For all indices $0 < i < s$, we have $\dim K_i \leq \frac{\dim F_i}{n} m$ (resp. $<$). Recall that the sequence r_i is decreasing. So the above formula shows that

$$-\mu(\lambda, K) \leq \frac{m}{n} \left(\sum_{i=1}^{s-1} (r_i - r_{i+1}) \dim F_i + r_s n \right) \text{ (resp. } <).$$

The last sum is $\sum_{i=1}^{s} r_i \dim H_i = 0$. That implies $\mu(\lambda, K) \geq 0$ (resp. $>$), and then K must be semi-stable (resp. stable) with respect to the group $\mathrm{SL}(H)$.

Conversely, let $H_1 \subset H$ be a vector sub-space of dimension $0 < p < n$. Let H_2 be a vector sub-space such that $H_1 \oplus H_2 = H$. If K is semi-stable (resp. stable) for all one parameter sub-groups of $\mathrm{SL}(H)$, it is semi-stable (resp. stable) for the one parameter sub-group defined in the direct sum $H = H_1 \oplus H_2$ by the matrix

$$\begin{pmatrix} t^{n-p}\mathrm{id}_{H_1} & 0 \\ 0 & t^{n-p}\mathrm{id}_{H_2} \end{pmatrix}.$$

Let $K_1 = (H_1 \otimes V) \cap K$. We have $-\mu(\lambda, K) = n \dim K_1 - p \dim K$. So

$$\frac{\dim K_1}{p} \leq \frac{\dim K}{n}.$$

In the stability case, we evidently have a strict inequality. □

7

The Construction of $\mathbf{M}(r, d)$

Let X be a smooth connected projective algebraic curve. Two semi-stable bundles on X are said to be S-equivalent if their associated Jordan-Hölder gradings are isomorphic. In this chapter we aim to give the set of equivalence classes of semi-stable vector bundles the structure of a projective algebraic variety. We shall show that the semi-stable points in the variety $\mathbf{Hilb}^P(H \otimes \mathcal{O}_X)$, suitably embedded in a projective space, under the $\mathrm{SL}(H)$ action can be identified with the points corresponding to semi-stable sheaves. The method which follows here is that of C. Simpson [58]. As distinct from the original construction of Mumford and Seshadri [55], Simpson's construction extends without difficulty to moduli spaces of semi-stable sheaves on singular projective curves as well as to higher dimensional projective varieties. Of course there are additional technical difficulties in the latter case (cf. Part II for the case of surfaces). For simplicity we shall limit ourselves to the case of smooth curves. The classical method of Mumford and Seshadri is described in [54] where one can also find variants concerning parabolic structures and level structures. One could also consult Oesterlé's article [51].

7.1 The semi-stability criterion

We can rephrase the notion of semi-stability in terms of sections. Fix a point $a \in X$ and, for any coherent sheaf F on X, set

$$F(n) = F \otimes \mathcal{O}_X(na).$$

Proposition 7.1.1 *Let $Coh(r,d)$ be the family of coherent sheaves on X of rank r and degree d. There exists an integer $N(r,d)$ such that for $n \geq N(r,d)$ and $F \in Coh(r,d)$ the following are equivalent:*

(i) the sheaf F is locally-free and semi-stable;

(ii) for all coherent sub-sheaves F' of F with rank r' we have

$$h^0(F'(n)) \leq \frac{r'}{r} h^0(F(n)).$$

We shall prove this after the following lemma.[1]

Lemma 7.1.2 *For any semi-stable bundle F on X of rank r and slope μ we have*

$$\frac{h^0(F)}{r} \leq [\mu+1]_+ .$$

Proof If $\mu < 0$ we obviously have $h^0(F) = 0$ and the lemma is true. Now suppose that $\mu \geq 0$ and induct on the degree. We assume that the lemma has been proved in degree less than d. We have an exact sequence

$$0 \longrightarrow F(-1) \longrightarrow F \longrightarrow F(a) \longrightarrow 0,$$

where $F(a) = F_a \otimes_{\mathcal{O}_a} \mathbb{C}$ is the fibre of F at the point a. This shows that $\mu(F) = \mu(F(-1)) + 1$ and $h^0(F) \leq h^0(F(-1)) + r$. The induction hypothesis applied to $F(-1)$ shows that

$$\frac{h^0(F(-1))}{r} \leq \mu.$$

It follows that $h^0(F)/r \leq \mu + 1$. □

Proof of Proposition 7.1.1 (i) $\Rightarrow$ (ii): Let F be a semi-stable bundle of rank r and degree d, and let $F' \subset F$ be a locally-free sub-sheaf. We can assume, by taking a larger F', that F/F' is locally-free. Let r' be the rank, μ'_i the Harder-Narasimhan sequence of F' and r'_i the sequence of ranks of the Harder-Narasimhan grading. By Lemma 7.1.2, we have, setting $\nu = \mu_{\min}(F')$,

$$\begin{aligned}\frac{h^0(F'(n))}{r'} &\leq \sum_i \frac{r'_i}{r'}[\mu'_i + n + 1]_+ \\ &\leq \left(1 - \frac{1}{r}\right)[\mu + n + 1]_+ + \frac{1}{r}[\nu + n + 1]_+.\end{aligned}$$

[1] For any real number x we put $[x]_+ = \sup(x, 0)$.

Choose an integer $a < \mu - gr$. For $\nu \leq a$ and $a + n + 1 \geq 0$ the right hand side is bounded above by $\mu + n + 1 - g$ because then $\mu + n + 1 \geq 0$ and $a/r + 1 + \mu(1 - 1/r) < \mu + 1 - g$. Hence,

$$\frac{h^0(F'(n))}{r'} < \mu + n + 1 - g.$$

We can then take $N(r,d) \geq -(a+1)$ and such that for $n \geq N(r,d)$ we have $H^1(F(n)) = 0$, which is possible because the family $\mathcal{S}(r,d)$ of semi-stable sheaves of rank r and degree d is bounded (Theorem 5.5.1). Then

$$\frac{\chi(F(n))}{r} = \frac{h^0(F(n))}{r} = n + \mu + 1 - g.$$

Consequently, for sub-sheaves $F' \subset F$ such that $\mu_{\min}(F') \leq a$, we have the inequality

$$\frac{h^0(F'(n))}{r'} < \frac{H^0(F(n))}{r}.$$

The sub-bundles $F' \subset F$ for which $\mu_{\min}(F') > a$ also form a bounded family. Indeed, from what we have seen in section 5.5, the semi-stable bundles $F \in \mathcal{S}(r,d)$ are quotients of a well defined bundle E and the quotient bundles of some bundle F of rank r and degree d form a bounded family since they appear amongst the quotients of E. The same follows for the family of sub-bundles $F' \subset F$ of fixed rank and degree. The family we are considering is then a finite union of bounded families of vector bundles and so is itself bounded.

We now choose $N(r,d)$ large enough so that for all $n \geq N(r,d)$ we have $h^1(F'(n)) = 0$ and so $h^0(F'(n)) = \chi(F'(n))$. For these sub-sheaves the desired inequality is therefore a trivial consequence of the definition of semi-stability.

(ii) $\Rightarrow$ (i): the difficulty here lies in the fact that the family $\mathcal{C}oh(r,d)$ is not bounded. We can assume that $\mu + N(r,d) \geq g$. A torsion sheaf must have sections if it is non-zero and so we see that the hypothesis implies that F is torsion-free and so locally-free. Let F'' be a semi-stable quotient of F of rank r'' and slope $\mu'' \leq \mu$. By Lemma 7.1.2 we have

$$\mu + n + 1 - g \leq \frac{h^0(F(n))}{r} \leq [\mu'' + 1 + n]_+$$

from which it follows that $\mu - g \leq \mu''$. This property and Theorem 5.6.1 imply that the family $\mathcal{T}(r,d)$ of sheaves F satisfying condition (ii) for at least one integer n is bounded; this is the same for the family of quotients F'' we are considering since the slope μ'' is bounded. If $N(r,d)$

is taken to be large enough we can then assume, by Serre's Theorem B, that for n larger than $N(r,d)$, we have $h^0(F(n)) = \chi(F(n))$ and $h^0(F''(n)) = \chi(F''(n))$. It follows from condition (ii) that

$$\frac{\chi(F''(n))}{r''} \geq \frac{\chi(F(n))}{r}$$

which is equivalent to $\mu'' \geq \mu$. Thus, F is semi-stable which completes the proof. □

Proposition 7.1.3 *There exists an integer $N(r,d)$ which satisfies the conditions of Proposition 7.1.1 and such that for any $n \geq N(r,d)$, any semi-stable bundle F of rank r and degree d and any non-zero coherent sub-sheaf F' of F of rank r', the following statements are equivalent:*

(a) $\mu(F') = \mu(F)$;

(b) $\dfrac{h^0(F'(n))}{r'} = \dfrac{h^0(F(n))}{r}$.

Proof If condition (a) holds then the sheaf F' must be semi-stable and the family of sub-sheaves F' thus obtained is bounded. It follows that if $N(r,d)$ is sufficiently large and $n \geq N(r,d)$ then we have $\chi(F'(n)) = h^0(F'(n))$ and so, as we have already seen, the same holds for F. Hence, condition (b) holds. For the other way round, if (b) holds for $n \geq N(r,d)$ then we must have $\mu_{\min}(F') > a$, where a is the integer determined in the proof of Proposition 7.1.1. Hence, the family of sub-sheaves F' is still bounded. It follows that if $N(r,d)$ is chosen to be sufficiently large then $\chi(F'(n)) = h^0(F'(n))$ and the same for F. In particular, we have $\mu(F'(n)) = \mu(F(n))$ from which we get (a). □

7.2 Construction of the moduli space

We consider the contravariant functor $\underline{\mathbf{M}}(r,d)$ which associates to an algebraic variety S the set $\underline{\mathbf{M}}(r,d)(S)$ of isomorphism classes of algebraic vector bundles F of rank r and degree d on $S \times X$ such that the bundle $F(s)$ is semi-stable of degree d for all $s \in S$.

Theorem 7.2.1 *There exists a coarse moduli space $\mathbf{M}(r,d)$ for the functor $\underline{\mathbf{M}}(r,d)$. This is a projective variety whose closed points correspond to S-equivalence classes of semi-stable bundles of rank r and degree d. There exists an open subset $\mathbf{M}^s(r,d)$ of $\mathbf{M}(r,d)$ whose points represent the isomorphism classes of stable bundles.*

By *coarse moduli* we mean that there exists a functorial morphism $\varphi : \underline{\mathbf{M}}(r,d)(S) \to \mathrm{Mor}(S, \mathbf{M}(r,d))$ which is universal in the following sense: for each algebraic variety N and functorial morphism $\psi : \underline{\mathbf{M}}(r,d)(S) \to \mathrm{Mor}(S,N)$ there exists one and only one morphism $f : \mathbf{M}(r,d) \to N$ making the following commute:

$$\begin{array}{ccc} \underline{\mathbf{M}}(r,d)(S) & \xrightarrow{\ \varphi\ } & \mathrm{Mor}(S, \mathbf{M}(r,d)) \\ & \searrow^{\psi} & \downarrow f \\ & & \mathrm{Mor}(S,N) \end{array}$$

The pair $(\mathbf{M}(r,d), \varphi)$ is uniquely determined by this condition. The morphism φ induces a map of sets $\mathcal{S}(r,d) = \underline{\mathbf{M}}(r,d)(\text{point})$ to $\mathbf{M}(r,d)$. The second assertion means that this map surjects and the fibres consist of S-equivalence classes of semi-stable bundles.

The construction of $\mathbf{M}(r,d)$ and the coarse moduli property will be described in this section. The description of the points of $\mathbf{M}(r,d)$ will be given in section 7.4 after some remarks about extensions of coherent sheaves.

Proof For each integer ν, we have $\mathbf{M}(r,d) \simeq \mathbf{M}(r, r\nu + d)$ and, by Propositions 7.1.1 and 7.1.3, we can assume that d is sufficiently large so that the following properties hold.

(i) Every semi-stable bundle F of rank r and degree d is generated by its sections and we have $H^1(F) = 0$, and the same thing holds for non-zero coherent sub-sheaves $F' \subset F$ with $\mu(F') = \mu(F)$.

(ii) A coherent sheaf $F \in \mathrm{Coh}(r,d)$ is locally-free and semi-stable if and only if for each non-zero coherent sub-sheaf $F' \subset F$ of rank r', we have

$$h^0(F') \leq \frac{r'}{r} h^0(F).$$

(iii) Given a semi-stable bundle F of rank r and degree d, a non-zero coherent sub-sheaf has slope $\mu(F') = \mu(F)$ if and only if

$$\frac{h^0(F')}{r} = \frac{h^0(F)}{r}.$$

We may further assume that $\mu = d/r > 2g - 1$. Let H be a vector space of dimension $\chi = d + r(1-g)$, the degree d having been chosen as above. We consider the Hilbert scheme of coherent quotients F of $H \otimes \mathcal{O}_X$ of rank r and degree d. For each point $q \in \mathbf{Hilb}^{r,d}(H \otimes \mathcal{O}_X)$

we let F_q be the sheaf quotient $H \otimes \mathcal{O}_X$ defined by this point. Let $A_n = \Gamma(\mathcal{O}_X(n))$. There exists an integer n as large as we please so that for each $q \in \mathbf{Hilb}^{r,d}(H \otimes \mathcal{O}_X)$, the vector space $H^0(F_q(n))$ is a dimension $rn + \chi$ quotient of the vector space $H \otimes A_n$. By Corollary 4.4.10, the morphism

$$\begin{array}{rcl} q & \mapsto & H^0(F_q(n)) : \\ \varphi_n : \mathbf{Hilb}^{r,d}(H \otimes \mathcal{O}_X) & \longrightarrow & \mathbf{Grass}^{rn+\chi}(H \otimes A_n), \end{array}$$

of $\mathbf{Hilb}^{r,d}(H \otimes \mathcal{O}_X)$ to the Grassmannian of quotient vector spaces of $H \otimes A_n$ of dimension $rn + \chi$ is an embedding for a suitable choice of n. This embedding φ_n is equivariant for the action of $\mathrm{SL}(H)$ which allows us to study the semi-stable points under the action of $\mathrm{SL}(H)$. In the following, the Hilbert polynomial P of a coherent sheaf F of rank r and Euler-Poincaré characteristic χ is the polynomial $\chi(F(m)) = rm + \chi$ relative to the ample line bundle $\mathcal{O}_X(a)$.

Lemma 7.2.2 *For $q \in \mathbf{Hilb}^{r,d}(H \otimes \mathcal{O}_X)$ and a suitable choice of n the following are equivalent:*

(1) the point q is semi-stable (respectively stable) for the action of $SL(H)$;

(2) for each non-zero proper vector sub-space $H' \subset H$, we have

$$\frac{P'}{\dim H'} \geq \frac{P}{\dim H} \qquad \textit{(respectively >)}$$

where P' is the Hilbert polynomial of the sub-sheaf F' of $F = F_q$ generated by H'.

In this statement, the linear polynomials are ordered lexicographically beginning with the dominant term. We need the notion of a bounded set of isomorphism classes of coherent sheaves on X. A set $\mathcal{C}$ of isomorphism classes of coherent sheaves on X is said to be *bounded* if there exist an algebraic variety S and a coherent sheaf $\mathcal{G}$ on $S \times X$ such that the set of isomorphism classes of sheaves $\mathcal{G}(s)$ for $s \in S$ contains $\mathcal{C}$. As in Lemma 4.4.8, if we need to we can replace S by a finite disjoint union S' of sub-varieties S_i over which $\mathcal{G}$ is flat. Moreover, we can assume in the above definition that $\mathcal{G}$ is S-flat. This implies that the family of Hilbert polynomials of sheaves contained in $\mathcal{C}$ is finite by Corollary 4.2.11.

Lemma 7.2.3 *If $\mathcal{C}$ and $\mathcal{C}'$ are two bounded families of coherent sheaves on X then the sets of kernels, images and cokernels of maps $G \to G'$, for $G \in \mathcal{C}$ and $G' \in \mathcal{C}'$, are also bounded.*

Proof Consider the coherent sheaves $\mathcal{G}$ and $\mathcal{G}'$ on $S \times X$ assumed S-flat and defining $\mathcal{C}$ and $\mathcal{C}'$. We can further assume, by changing S if necessary[1], that there is a map $f : \mathcal{G} \to \mathcal{G}'$ such that the induced maps $f(s)$ for $s \in S$ give rise to all the given maps. So, if we consider the cokernel $\mathcal{Q} = \operatorname{coker} f$, then the set of isomorphism classes of the $\mathcal{Q}(s)$ contains all of the cokernels required in the lemma. For the images we start by changing bases via $j : S' \to S$ so that the sheaf $j^*(\mathcal{Q})$ becomes S'-flat, as we have just explained, so that we are reduced to the case where $\mathcal{Q}$ is S-flat. In this case we have, for each $s \in S$, the identity $\operatorname{Im} f(s) = (\operatorname{Im} f)(s)$ and $\ker f(s) = (\ker f)(s)$ from which we obtain the lemma. □

Proof of Lemma 7.2.2 By Lemma 7.2.3 the families of sheaves F and F' which occur in the lemma are bounded as is the family of kernels of the epimorphisms $H' \otimes \mathcal{O}_X \to F'$. Therefore, we can assume, for $n \geq N(r,d)$, that $H^1(F(n)) = 0$, $H^1(F'(n)) = 0$ and that the canonical linear map $H' \otimes A_n \to H^0(F'(n))$ surjects. To say that the point q is semi-stable for the embedding defined by φ_n means, recalling chapter 6, that the point $\varphi_n(q)$ is a semi-stable point of the Grassmannian $\mathbf{Grass}^{P(n)}(H \otimes A_n)$, which allows us to write the inequality

$$\frac{h^0(F'(n))}{\dim H'} \geq \frac{h^0(F(n))}{\dim H}$$

for all H'. But the family $\mathcal{P}$ of polynomials which appear in these inequalities is finite. So we can further assume that, for each $n \geq N(r,d)$, the map $\mathcal{P} \to \mathbf{Q} : P \mapsto P(n)$ is strictly increasing. Thus, for such n, the inequality above is equivalent to

$$\frac{P'}{\dim H'} \geq \frac{P}{\dim H}.$$

This proves Lemma 7.2.2. □

Lemma 7.2.4 *For a suitable choice of n and for all points*

$$q \in \mathbf{Hilb}^{r,d}(H \otimes \mathcal{O}_X)$$

the following are equivalent:

(a) *the sheaf F_q is semi-stable and the linear map $H \to H^0(F_q)$ induced by the canonical projection is an isomorphism;*

[1] This point is a little delicate: we need to consider an algebraic variety $T \to S$ whose fibres over $s \in S$ are the vector spaces $\operatorname{Hom}(\mathcal{G}(s), \mathcal{G}'(s))$.

(b) the point q is semi-stable for the action of $\mathrm{SL}(H)$.

Proof (1) $\Rightarrow$ (2): Let H' be a non-zero vector sub-space of H and let F' be the sub-sheaf of $F = F_q$ generated by H'. Let P' denote the Hilbert polynomial of F'. Since the point q is semi-stable we have, for a suitable choice of d,

$$\frac{h^0(F')}{r'} \leq \frac{h^0(F)}{r}.$$

On the other hand, the commuting diagram

$$\begin{array}{ccc} H' & \longrightarrow & H^0(F') \\ \downarrow & & \downarrow \\ H & \xrightarrow{\sim} & H^0(F) \end{array}$$

shows that $\dim H' \leq h^0(F')$ and it follows that $\dim H'/r' \leq \dim H/r$. Moreover, we get equality only if $h^0(F')/r' = h^0(F)/r$, that is, for a suitable choice of d, if $\mu(F') = \mu(F)$. This certainly implies

$$\frac{P'}{\dim H'} \geq \frac{P}{\dim H}.$$

(2) $\Rightarrow$ (1): We first show that q is semi-stable under the $\mathrm{SL}(H)$ action and if $\mu = d/r > 2g-1$ then we have $H^1(F_q) = 0$. For all coherent $\mathcal{O}_X$-modules, we have, by Grothendieck-Serre duality, $\mathrm{Hom}(F, \omega_X) = (H^1(F))^*$. Suppose that there exists a non-zero map $F_q \to \omega_X$. The image is an invertible sheaf F''. Let $H'' \hookrightarrow H^0(F'') \hookrightarrow H^0(\omega_X)$ be the image of H. Let P'' denote the Hilbert polynomial of F''. By the semi-stability criterion (Lemma 7.2.2), written in terms of quotients, we have $H'' \neq 0$ and

$$\frac{P}{\dim H} \geq \frac{P''}{\dim H''}.$$

The coefficient of the dominant term gives $\dim H/r \leq \dim H''/r''$ which implies $\chi/r \leq h^0(\omega_X) = g$ and so $\mu \leq 2g-1$. This contradicts the choice of d.

It follows that $h^0(F_q) = \chi$. We now show that the canonical linear map $H \to H^0(F_q)$ injects. If not, it has a non-zero kernel H' and the sub-sheaf generated by H' vanishes. Then $P' = 0$, which contradicts the semi-stability condition. Hence, the canonical linear map $H \to H^0(F_q)$ is an isomorphism.

To see that $F = F_q$ is a semi-stable bundle, let F' be a coherent sub-sheaf of $F = F_q$. We must show that if F' is non-zero then it has rank at least 1 and that the inequality

$$\frac{h^0(F')}{r'} \leq \frac{h^0(F)}{r}$$

holds. We can certainly assume that $h^0(F') \neq 0$ because if not, the result is trivial. We consider the sub-space $H' \subset H$ given by $H' = H^0(F')$ and apply the semi-stability criterion. Let F'' be the sub-sheaf of F generated by H' and let r'' be its rank and P'' its Hilbert polynomial. By Lemma 7.2.2, we have

$$\frac{P''}{\dim H'} \geq \frac{P}{\dim H}.$$

So, taking the coefficient of the dominant term, $r''/\dim H' \geq r/\dim H$. Since $r' \geq r''$, this means that $r'/\dim H' \geq r/\dim H$. It follows that r' is non-zero and that

$$\frac{h^0(F')}{r'} \leq \frac{h^0(F)}{r}.$$

By condition (ii), the bundle F is semi-stable. □

Lemma 7.2.5 *Let $q \in \mathbf{Hilb}^{r,d}(H \otimes \mathcal{O}_X)$ be a semi-stable point under the $\mathrm{SL}(H)$ action. A coherent sub-module $F' \subset F_q$ of rank $r' > 0$ with Hilbert polynomial P' and such that $\mu(F') = \mu$ is generated by a non-zero vector sub-space $H' \subset H$ such that*

$$\frac{P'}{\dim H'} = \frac{P}{\dim H}.$$

The correspondence between the vector sub-spaces H' and the sub-sheaves F' is a bijection.

Proof (1) Let H' be a non-zero vector sub-space of H such that the Hilbert polynomial of the sub-sheaf F' generated by H' satisfies the equality

$$\frac{P'}{\dim H'} = \frac{P}{\dim H}.$$

Then we have, from the dominant term, $\dim H'/r' = \dim H/r$ and so

$$\frac{\dim H'}{r'} \leq \frac{h^0(F')}{r'} \leq \frac{h^0(F)}{r}.$$

It follows that $h^0(F')/r' = h^0(F)/r$. By condition (iii), we then have $\mu(F') = \mu$. Furthermore, $H' = H^0(F')$.

(2) Let F' be a coherent sub-sheaf of $F = F_1$ such that $\mu(F') = \mu$. From condition (iii), we have $h^0(F')/r' = h^0(F)/r$. Consider the vector sub-space H' of H defined by $H' = H^0(F')$ and the sub-sheaf F'' generated by H' with rank r''. We have $h^0(F'') = h^0(F')$ and

$$\frac{h^0(F')}{r'} \leq \frac{h^0(F'')}{r''} \leq \frac{h^0(F)}{r}$$

and so $r'' = r$. This then implies that $F' = F''$, so F' is generated by H' and

$$\frac{P'}{\dim H'} = \frac{P}{\dim H}.$$

Hence, we obtain all of the sub-sheaves F' of F such that $\mu(F') = \mu$. Moreover, the construction carried out in (1) shows that if it comes from a sub-space H' satisfying the required conditions then for the sheaf F' generated by H' we have $H' = H^0(F')$. Hence, the correspondence is bijective. □

The coarse moduli property

Let Ω be the open subset of $\mathbf{Hilb}^{r,d}(H \otimes \mathcal{O}_X)$ of points semi-stable under the action of $\mathrm{SL}(H)$. By Theorem 6.4.4, we can then consider the projective variety $\mathbf{M}(r,d) = \Omega/\mathrm{SL}(H)$ of equivalence classes of quotients $H \otimes \mathcal{O}_X \to F$. If $\mathcal{F}$ is a rank r bundle on $S \times X$ giving rise to a family of semi-stable bundles on X of degree d parametrized by an algebraic variety S then the direct image sheaf $G = \mathrm{pr}_{1*}(\mathcal{F})$ under the first projection is a locally-free sheaf of rank χ on S due to the choice of degree (Corollary 4.2.12). There is a canonical $\mathrm{GL}(H)$-equivariant map from the frame bundle $\mathcal{R} = \underline{\mathrm{Isom}}(H \otimes \mathcal{O}_S, G)$ of G to $\mathbf{Hilb}^{r,d}(H \otimes \mathcal{O}_X)$ and, by Lemma 7.2.4, this map factorizes through the open set Ω so that we obtain a map $f_{\mathcal{F}} : S = \mathcal{R}/\mathrm{GL}(H) \to \mathbf{M}(r,d)$ which is obviously functorial in S.

We now check the coarse moduli property. Suppose that $\underline{\mathbf{M}}(r,d)(S) \to \mathrm{Mor}(S, \mathcal{N})$ is a map which is functorial in S. Let S be the open set Ω. Then S parametrizes a universal quotient bundle $\mathcal{F}$ equipped with an action of $\mathrm{SL}(H)$ covering the natural action on the open set Ω. This implies that the associated morphism $g_{\mathcal{F}} : \Omega \to \mathcal{N}$ is $\mathrm{SL}(H)$-equivariant. So, by the universal property of quotient, there exists one and only one map $\varphi : \mathbf{M}(r,d) \to \mathcal{N}$ such that $g_{\mathcal{F}} = \varphi \circ f_{\mathcal{F}}$.

The points of the open subset Ω^s of Ω which are stable for the $\mathrm{SL}(H)$ action are the points q which satisfy the following property: there is no proper vector sub-space $0 \subset H' \subset H$ such that, for the sub-sheaf F' generated by H', we have

$$\frac{P_{F'}}{\dim H'} = \frac{P}{\dim H}.$$

By Lemma 7.2.5 above, this means that the sheaf F_q is stable. The open set Ω^s is the inverse image of an open subset $\mathbf{M}^s(r,d)$ of $\mathbf{M}(r,d)$ by Proposition 6.4.7 and the morphism $\Omega^s \to \mathbf{M}^s(r,d)$ is a geometric quotient. By Theorem 5.5.2, the points of the open set $\mathbf{M}^s(r,d)$ can be identified with the isomorphism classes of stable bundles. This open set is obviously a coarse moduli space for the functor which associates to the algebraic variety S the set of isomorphism classes of families of stable bundles of rank r and degree d parametrized by S.

Semi-stability is an open condition

A variant of the above proof gives us the following result.

Proposition 7.2.6 *Let S be an algebraic variety and let $\mathcal{F}$ be an S-flat coherent sheaf on $S \times X$. Then the set of points $s \in S$ such that $\mathcal{F}(s)$ is semi-stable (respectively stable) is an open subset of S.*

Proof Let r be the rank of $\mathcal{F}$ and let d be the degree of the sheaf $\mathcal{F}(s)$ induced on the fibre over the point s. We can always assume that the degree is sufficiently large so that the sheaves $\mathcal{F}(s)$ are generated by their sections and such that $H^1(X, \mathcal{F}(s)) = 0$. Under these conditions, the direct image $\mathrm{pr}_{1*}(\mathcal{F})$ is locally-free of rank $\chi = d + r(1-g)$ and the choice of a local trivialization over an open subset U of S

$$H \otimes \mathcal{O}_U \simeq \mathrm{pr}_{1*}(\mathcal{F})$$

gives rise to a morphism $f : U \to \mathbf{Hilb}^{r,d}(H \otimes \mathcal{O}_X)$. The set of points $s \in U$ such that $\mathcal{F}(s)$ is semi-stable (respectively, stable) is, by Lemma 7.2.2, the inverse image of Ω (respectively, Ω^s) under the morphism f. This is therefore an open set. □

7.3 Extensions

Let F and G be two algebraic sheaves on an algebraic variety Y. Let $\mathrm{Ext}^q(F,G)$ denote the right derived functor of $G \mapsto \mathrm{Hom}(F,G)$. If

$G \xrightarrow{\epsilon} I^\bullet$ is an injective $\mathcal{O}_X$-module resolution of G then we have an isomorphism

$$\mathrm{Ext}^q(F,G) = H^q(\mathrm{Hom}(F,I^\bullet)).$$

When F is a locally-free sheaf, we have an isomorphism

$$\mathrm{Ext}^q(F,G) \simeq H^q(Y, \mathcal{H}om(F,G)).$$

By an *extension* of F by G we mean an exact sequence of algebraic sheaves $0 \to G \xrightarrow{i} E \xrightarrow{p} F \to 0$. We say that E is an *extension* of F by G. Given two extensions E and E' of F by G, a map of extensions is a map $f : E \to E'$ making the diagram

$$\begin{array}{ccccccccc} 0 & \longrightarrow & G & \longrightarrow & E & \longrightarrow & F & \longrightarrow & 0 \\ & & {\scriptstyle \mathrm{id}_G}\downarrow & & \downarrow{\scriptstyle f} & & \downarrow{\scriptstyle \mathrm{id}_F} & & \\ 0 & \longrightarrow & G & \longrightarrow & E' & \longrightarrow & F & \longrightarrow & 0 \end{array}$$

commute. We let $\mathbf{Ext}(F,G)$ denote the set of isomorphism classes of extensions of F by G. The group $\mathrm{Aut}(F) \times \mathrm{Aut}(G)$ acts naturally on the set $\mathbf{Ext}(F,G)$; in each orbit the isomorphism class of E remains unchanged.

Lemma 7.3.1 *We have a canonical isomorphism*

$$\mathbf{Ext}(F,G) \simeq \mathrm{Ext}^1(F,G)$$

compatible with the action of the group $\mathrm{Aut}(F) \times \mathrm{Aut}(G)$.

Proof Given an extension $0 \to G \xrightarrow{i} E \xrightarrow{p} F \to 0$, we can associate to it a long exact sequence given by the derived functors of $\mathrm{Hom}(F, -)$, and so we obtain a boundary homomorphism $\partial : \mathrm{Hom}(F,F) \to \mathrm{Ext}^1(F,G)$. Set $\omega = \partial(\mathrm{id}_F)$. This element only depends on the isomorphism class of the given extension. So we have a map

$$\Phi : \mathbf{Ext}(F,G) \longrightarrow \mathrm{Ext}^1(F,G).$$

Conversely, consider an element $\omega \in \mathrm{Ext}^1(F,G)$. For a given injective resolution $G \xrightarrow{\epsilon} I^\bullet$ we can represent the class ω by a map $w : F \to I^1$ such that $dw = 0$, where d denotes the differential in the complex $I^\bullet$.

We then obtain an exact sequence of complexes

$$\begin{array}{ccccccccc} 0 & \longrightarrow & I^2 & \longrightarrow & I^2 & \longrightarrow & 0 & & \\ & & {\scriptstyle d}\uparrow & & {\scriptstyle d}\uparrow & & & & \\ 0 & \longrightarrow & I^1 & \longrightarrow & I^1 & \longrightarrow & 0 & & \\ & & {\scriptstyle d}\uparrow & & \uparrow{\scriptstyle (d,w)} & & \uparrow & & \\ 0 & \longrightarrow & I^0 & \longrightarrow & I^0 \oplus F & \longrightarrow & F & \longrightarrow & 0 \end{array}$$

Let E_w be the kernel of the map (d, w). The exact sequence above gives rise to an extension $0 \to G \xrightarrow{i} E_w \xrightarrow{p} F \to 0$. If we replace w by another representative w' then we have $w' = w + dg$ with $g \in \mathrm{Hom}(F, I^0)$. We have a commuting diagram

$$\begin{array}{ccc} I^0 \oplus F & \xrightarrow{(d,w')} & I^1 \\ {\scriptstyle \varphi}\downarrow & & \downarrow{\scriptstyle \mathrm{id}} \\ I^0 \oplus F & \xrightarrow{(d,w)} & I^1 \end{array}$$

where φ is given by the matrix $\begin{pmatrix} \mathrm{id} & g \\ 0 & \mathrm{id} \end{pmatrix}$. Hence, we obtain an isomorphism $\varphi : E_{w'} \to E_w$ compatible with the associated extensions. We then obtain a map

$$\Psi : \mathrm{Ext}^1(F, G) \longrightarrow \mathbf{Ext}(F, G).$$

It remains to show that Ψ is the inverse of Φ. Let E be an extension of F by G. Since I^0 is an injective $\mathcal{O}_Y$-module, the map $G \xrightarrow{\epsilon} I^0$ extends to a map $\eta : E \to I^0$ and the map $d\eta : E \to I^1$ vanishes on G so that it factors through F to give a map $F \to I^1$ which we denote by $-w$. We have that $d\eta + wp = 0$ and $dw = 0$. Then we have an exact sequence of complexes

$$\begin{array}{ccccccccc} & & {\scriptstyle d}\uparrow & & {\scriptstyle d}\uparrow & & & & \\ 0 & \longrightarrow & I^1 & \longrightarrow & I^1 & \longrightarrow & 0 & & \\ & & {\scriptstyle d}\uparrow & & {\scriptstyle (d,w)}\uparrow & & \uparrow & & \\ 0 & \longrightarrow & I^0 & \longrightarrow & I^0 \oplus F & \longrightarrow & F & \longrightarrow & 0 \\ & & \uparrow & & {\scriptstyle (\eta,p)}\uparrow & & {\scriptstyle \mathrm{id}}\uparrow & & \\ 0 & \longrightarrow & G & \longrightarrow & E & \longrightarrow & F & \longrightarrow & 0 \\ & & \uparrow & & \uparrow & & \uparrow & & \\ & & 0 & & 0 & & 0 & & \end{array}$$

If we let $J^\bullet$ denote the resolution of E given by the above diagram then we can compute $\omega = \partial(\mathrm{id}_F)$ via the Snake Lemma and the exact sequence of complexes

$$0 \longrightarrow \mathrm{Hom}(F, I^\bullet) \longrightarrow \mathrm{Hom}(F, J^\bullet) \longrightarrow \mathrm{Hom}(F, F) \longrightarrow 0.$$

We find that the image under Φ of the above extension class is just the class ω of w in $\mathrm{Ext}^1(F, G)$. It is clear in the construction of Ψ given above that the extension associated to ω under the map Ψ is just the original extension. So $\Psi \circ \Phi = \mathrm{id}$.

For the converse, if we have a class $\omega = [w]$ in $\mathrm{Ext}^1(F, G)$ then to define $\eta : E_w \to I^0$ we can choose the first factor of the inclusion $E_w \hookrightarrow I^0 \oplus F$. Since E_w is the kernel of (d, w) we have the formula $d\eta + pw = 0$ which proves that the associated map $F \to I^1$ is just w. It follows that $\Phi \circ \Psi = \mathrm{id}$. □

The universal extension

Suppose that Y is a projective variety and that F and G are locally-free sheaves. Then $\mathcal{V} = \mathrm{Ext}^1(F, G)$ is a finite dimensional vector space. Over the product $\mathcal{V} \times Y$ we consider the vector bundles $\mathbf{F} = \mathrm{pr}_2^*(F)$ and $\mathbf{G} = \mathrm{pr}_2^*(G)$, where pr_2 is the projection to the second factor. It follows from Proposition 4.2.2 that the space $\mathrm{Ext}^1(\mathbf{F}, \mathbf{G})$ is the vector space of regular maps $\mathcal{V} \to \mathcal{V}$. In particular, this vector space contains the distinguished element corresponding to the identity of $\mathcal{V}$ to which we can associate an extension on $\mathcal{V} \times Y$,

$$0 \longrightarrow \mathbf{G} \longrightarrow \mathbf{E} \longrightarrow \mathbf{F} \longrightarrow \mathbf{0},$$

called the *universal extension*. At each point $\omega \in \mathcal{V}$, the extension associated to $\{\omega\} \times Y$ is just the extension $\Psi(\omega)$ constructed above.

7.4 The points of $\mathbf{M}(r, d)$

We shall show that if F is a semi-stable bundle of rank r and degree d then the bundles F and $\mathrm{gr}(F)$ define the same point in $\mathbf{M}(r, d)$. It suffices to show that if F_1 is a sub-bundle of F of slope μ then the bundles F and $F_1 \oplus F/F_1$ define the same point in $\mathbf{M}(r, d)$. Now, there exists a family of extensions $(F_\eta)_{\eta \in \mathrm{Ext}^1(F/F_1, F_1)}$ which define a map

$$\mathrm{Ext}^1(F/F_1, F_1) \longrightarrow \mathbf{M}(r, d).$$

For $t \neq 0$, we have $F_{t\eta} \simeq F_\eta$ which shows that this map is constant on straight lines through the origin. Hence, it is constant; for $\eta = 0$ we certainly obtain the point given by the direct sum $F_1 \oplus F/F_1$.

It remains to see that if two polystable bundles (that is, direct sums of stable bundles of the same slope) are not isomorphic then they define distinct points of $\mathbf{M}(r, d)$. Let F be such a polystable bundle. By Mumford's property of quotients, it suffices to prove that the orbit of the point $q \in \Omega$ associated to (F, α), where $\alpha : H \simeq H^0(F)$ is an isomorphism, is closed in Ω. Let q_∞ be a point in the support of the orbit corresponding to an isomorphism class of the pair $(F_\infty, \alpha_\infty)$, where F_∞ is a semi-stable bundle and $\alpha_\infty : H \simeq H^0(F_\infty)$ is an isomorphism. Let

$$F = \bigoplus_i F_i^{\rho_i}$$

be a decomposition of F into a direct sum of stable bundles. By semi-continuity (Corollary 4.2.11) we have that $\dim \operatorname{Hom}(F_i, F_\infty) \geq \rho_i$. It follows that there is an embedding

$$\bigoplus_i F_i^{\rho_i} \hookrightarrow F_\infty$$

and hence an isomorphism $F \simeq F_\infty$. These two bundles then define points in the same orbit. This proves that the orbit of q is closed and completes the proof of Theorem 7.2.1. $\square$

8

Study of $\mathbf{M}(r, d)$

As in the previous chapter, X will be a smooth projective curve of genus g. The main object of this chapter is to study the open subset of smooth points in the moduli space $\mathbf{M}(r, d)$ of S-equivalence classes of semi-stable bundles of rank r and degree d on X as well as the irreducibility of the moduli space. We show that this moduli space is smooth at the points representing stable bundles. In fact, we know that there are no other smooth points except in the particular cases $g = r = 2$ and even d [48]. The essential point is the smoothness criterion for points in the Hilbert scheme. This criterion, due to Grothendieck [29], reduces our study of smoothness to the problem of extending flat families of coherent sheaves parametrized by fat points. We also examine the conditions under which the moduli spaces are not empty. As we do this we will obtain the existence of a universal bundle when r and d are coprime and so, in this case, the moduli space $\mathbf{M}(r, d)$ is an irreducible smooth projective variety and there exists a universal bundle over $\mathbf{M}(r, d) \times X$. In fact, we can show that this is the only case where there exists a universal bundle: this is Ramanan's Theorem [52, 16].

8.1 The smoothness criterion

Consider an algebraic variety Y and let $A = \mathcal{O}_{Y,a}$ denote the local algebra of Y at a point a. Let $\mathfrak{m}$ be the maximal ideal of A and $T_aY = (\mathfrak{m}/\mathfrak{m}^2)^*$ the Zariski tangent space. We let $D = \mathrm{Spec}(\mathbb{C}[\epsilon]/(\epsilon^2))$ denote the variety supported at a point associated to the algebra of dual numbers.

Lemma 8.1.1 *Let* $\mathrm{Mor}_a(D,Y)$ *denote the set of morphisms* $f : D \to Y$ *with image supported at* $a \in Y$*. Then there is a canonical isomorphism*

$$\mathrm{Mor}_a(D,Y) \simeq T_aY.$$

Proof Let $f \in \mathrm{Mor}_a(D,Y)$. The Zariski tangent space of D is canonically isomorphic to $\mathbb{C}$ and consequently the differential of such a morphism defines a tangent vector $v \in T_aY$. Then we obtain a map

$$\mathrm{Mor}_a(D,Y) \longrightarrow T_aY$$

which is functorial in Y. Choosing a local model for the variety Y in a neighbourhood of the point a we are reduced to showing that this map is an isomorphism when $Y = \mathbb{C}^n$. So it suffices to verify the result when $Y = \mathbb{C}$, but then it is obvious. □

Definition 8.1.2 *By the* cone *on* Y *at* a *we mean the algebraic variety*

$$C = \mathrm{Spec}(\bigoplus_{i \geq 0} (\mathfrak{m}^i/\mathfrak{m}^{i+1})).$$

This algebraic variety is a cone and we use 0 to denote the vertex. The Zariski tangent space of C can be identified with that of Y at a. Moreover, we have a canonical embedding $(C,0) \hookrightarrow (T_aY,0)$. Indeed, let $\mathrm{Sym}(\mathfrak{m}/\mathfrak{m}^2) = \bigoplus_{i\geq 0} S^i(\mathfrak{m}/\mathfrak{m}^2)$ be the polynomial algebra of the tangent space T_aY. Then we have a canonical surjection

$$\bigoplus_{i\geq 0} S^i(\mathfrak{m}/\mathfrak{m}^2) \longrightarrow \bigoplus_{i\geq 0}(\mathfrak{m}^i/\mathfrak{m}^{i+1})$$

which is compatible with the grading. It follows that we get an embedding of the algebraic varieties which represent these algebras.

Lemma 8.1.3 *We have* $\dim_a Y = \dim_0 C$.

Proof Let $d = \dim_a Y$. We know (cf. Atiyah and MacDonald [4]) that, for n sufficiently large, the function $n \mapsto \dim_{\mathbb{C}}(A/\mathfrak{m}^n)$ is a polynomial in n of degree d: this is the Hilbert-Samuel polynomial

$$\dim_{\mathbb{C}}(A/\mathfrak{m}^n) = e\frac{n^d}{d!} + \text{lower order terms},$$

where e is the multiplicity. The same thing applies to the cone. But we find that the dimensions are the same for the cone. □

Proposition 8.1.4 *The variety Y is smooth at a if and only if the cone C is smooth at its vertex.*

Proof By definition, the variety Y is smooth at a if and only if $\dim_a Y = \dim T_a Y$ and the same for the cone. Since $(C, 0)$ and (Y, a) have the same Zariski tangent space, we obtain the result in view of Lemma 8.1.3. □

Corollary 8.1.5 *Let $T = T_a Y$ be the Zariski tangent space of Y at a. If the variety Y is smooth at a then the canonical embedding*

$$(C, 0) \longrightarrow (T, 0)$$

is an isomorphism.

Proof These two varieties are cones with vertices 0. At the origin, we have seen that they have the same dimension. Since $(T, 0)$ is integral, the map above is an isomorphism. □

Consider the category $\mathcal{A}$ of algebraic varieties of the form $\operatorname{Spec} B$, where B is an Artinian local finitely generated $\mathbb{C}$-algebra. The underlying topological space is reduced at the point representing the maximal ideal. Such a variety is called a *fat point.*

Theorem 8.1.6 *The following are equivalent:*

(i) the variety Y is smooth at a,
(ii) for each embedding of fat points $Z \hookrightarrow Z'$, any map $f \in \operatorname{Mor}_a(Z, Y)$ extends to Z'.

For the proof we need the notion of infinitesimal neighbourhood.

Definition 8.1.7 *The nth* infinitesimal neighbourhood *of a in Y is defined to be the fat point given by*

$$Y_n = \operatorname{Spec}(A/\mathfrak{m}^{n+1})$$

whose vertex will also be denoted by a.

Obviously, if $f : (Y, a) \to (Y', a')$ is a map of algebraic varieties such that $f(a) = a'$, and Y_n and Y'_n are the infinitesimal neighbourhoods of a in Y and of a' in Y', respectively, then we obtain a map $f_n : (Y_n, a) \to (Y'_n, a')$.

Lemma 8.1.8 *Suppose that Y is smooth at a of dimension d. Let $(z_1, \ldots, z_d)$ be a system of generators of $\mathfrak{m}$, $V = \mathbb{C}^d$ and $j : (Y, a) \to (V, 0)$ be the map of germs of the associated algebras. Then, for all n, the maps j induce an isomorphism on the infinitesimal neighbourhood of order n:*

$$j_n : (Y_n, a) \longrightarrow (V_n, 0).$$

Proof We induct on n. For $n = 1$ this means that j is étale at a. Suppose that we have proved the lemma up to order n. The variety Y_n has ideal $\mathfrak{m}^n/\mathfrak{m}^{n+1}$ in Y_{n+1} and V_n has ideal $S^n(\mathfrak{m}/\mathfrak{m}^2)$ in V_{n+1}. By Corollary 8.1.5, the canonical embedding $j : (C, 0) \to (T, 0)$ is an isomorphism; hence, the map $S^n(\mathfrak{m}/\mathfrak{m}^2) \to \mathfrak{m}^n/\mathfrak{m}^{n+1}$ is also an isomorphism. So $j_{n+1} : (Y_{n+1}, a) \to (V_{n+1}, 0)$ is an isomorphism. □

Proof of Theorem 8.1.6 (i) ⇒ (ii): The statement is obvious when we take Y to be the affine space $\mathbb{C}^d$. For the general case, suppose that Y is smooth and has dimension d at a. Choose a system of generators $(z_1, \ldots, z_d)$ of $\mathfrak{m}$ and set $V = \mathbb{C}^d$. Let $i : (Z, 0) \to (Z', 0)$ be an embedding of fat points. We consider an extension g' of $j \circ f = g$:

$$\begin{array}{ccc} (Z',0) & \xrightarrow{g'} & (V,0) \\ {\scriptstyle i}\uparrow & & \uparrow{\scriptstyle j} \\ (Z,0) & \xrightarrow[f]{} & (Y,a) \end{array}$$

For large enough n, the map f factors through the infinitesimal neighbourhood Y_n and the map g' factors through the infinitesimal neighbourhood V_n. Then we obtain the commuting diagram

$$\begin{array}{ccc} (Z',0) & \xrightarrow{g'_n} & (V_n,0) \\ {\scriptstyle i}\uparrow & & \uparrow{\scriptstyle j_n} \\ (Z,0) & \xrightarrow[f_n]{} & (Y_n,a) \end{array}$$

We know from Lemma 8.1.8 that the map inducing j_n on the infinitesimal neighbourhoods of order n is an isomorphism. We can then consider the map $f'_n = j_n^{-1} g'_n : (Z', 0) \to (Y_n, a)$ which extends f_n. The composite

$$f' : (Z', 0) \xrightarrow{f'_n} (Y_n, a) \hookrightarrow (Y, a)$$

produces the required extension of f.

(ii) $\Rightarrow$ (i): Let $T = T_aY$ be the Zariski tangent space of Y at a with base point the origin. Consider the infinitesimal neighbourhoods of $(T, 0)$ and (Y, a): to first order these neighbourhoods coincide. We then have maps $(T_1, 0) \simeq (Y_1, a) \hookrightarrow (Y, a)$ which extend, by (ii), to a map $(T_n, 0) \to (Y, a)$. Taking the cone, we obtain a map $\varphi_n : (T_n, 0) \to (C, 0)$. The composite

$$i_n : (T_n, 0) \longrightarrow (C, 0) \hookrightarrow (T, 0)$$

restricted to T_1 is just the canonical embedding. It follows that this map is also an embedding and therefore so too is φ_n. From this we have that

$$\dim_{\mathbb{C}} \mathcal{O}_{T_n,0} \leq \dim_{\mathbb{C}} \mathcal{O}_{C_n,0}.$$

From this and the Hilbert-Samuel polynomial we see that $\dim T \leq \dim_0 C = \dim_a Y$ and the inequality the other way round is obvious and so we have equality. Hence, Y is smooth at a and we have proved the theorem. □

8.2 The Hilbert scheme

Let E be a vector bundle on a smooth projective curve X of genus g and let $\mathbf{Hilb}^{r,d}(E)$ be the Hilbert scheme of coherent quotient sheaves of E of rank r and degree d.

Theorem 8.2.1 *Let $q \in \mathbf{Hilb}^{r,d}(E)$ be a point given by a coherent sheaf $F = E/G$.*

(i) The Zariski tangent space $T_q \mathbf{Hilb}^{r,d}(E)$ to the Hilbert scheme at q is isomorphic to $\mathrm{Hom}(G, F)$.

(ii) If $\mathrm{Ext}^1(G, F) = 0$ then the variety $\mathbf{Hilb}^{r,d}(E)$ is smooth in a neighbourhood of q.

In the case where we take E to be the trivial bundle $H \otimes \mathcal{O}_X$, we obtain the following corollary.

Corollary 8.2.2 *Let U be the open subset of $\mathbf{Hilb}^{r,d}(H \otimes \mathcal{O}_X)$ consisting of the points corresponding to semi-stable quotient sheaves F. Then U is smooth if $\mu = d/r \geq 2g - 1$.*

Proof Let q be a point of U defining a coherent quotient sheaf $F = E/G$. Then we have the exact sequence

$$\mathrm{Ext}^1(E, F) \longrightarrow \mathrm{Ext}^1(G, F) \longrightarrow 0$$

and $H^1(F) = 0$ which follows from our hypothesis and Serre duality since semi-stability gives $\mathrm{Hom}(F, \omega_X) = 0$. Since E is trivial, this implies that $\mathrm{Ext}^1(E, F) = 0$ and so $\mathrm{Ext}^1(G, F) = 0$. □

Proof of Theorem 8.2.1 Let S be an algebraic variety. Let $\mathbf{E} = E_S$ denote the pullback of E to $S \times X$. We first recall that, given an algebraic variety S, the set of morphisms $\mathrm{Mor}(S, \mathbf{Hilb}^{r,d}(E))$ can be identified with the set of coherent quotient sheaves $F = \mathbf{E}/G$ of $\mathbf{E}$ which are S-flat and of rank r and degree d over each point of S, and this isomorphism is functorial in S.

Now consider an embedding $S \hookrightarrow S'$ of fat points and set $\mathbf{E} = E_S$ and $\mathbf{E}' = E_{S'}$. Given an S-flat coherent quotient sheaf $F = \mathbf{E}/G$, by Lemma 8.1.1 and Theorem 8.1.6, we need to

(1) find sufficient conditions such that this sheaf can be extended to an S'-flat coherent quotient sheaf $\mathbf{E}'$ and,

(2) when this exists, to classify such extensions.

Suppose that $S = \mathrm{Spec}\, A$ and $S' = \mathrm{Spec}\, A'$, where A and A' are Artinian local algebras, and that S is given in S' by an ideal I of square zero. Let $F' = \mathbf{E}'/G$ be an S'-flat coherent sheaf which induces F over S; this condition means that the quotient $\mathbf{E} = \mathbf{E}' \otimes_{A'} A \to F' \otimes_{A'} A$ coincides with $\mathbf{E} \to F$. In terms of sub-sheaves, we can say that G is the image of $G' \otimes_{A'} A$ in $\mathbf{E} = \mathbf{E}' \otimes_{A'} A$ and the flatness of F' is equivalent to the injectivity of the map $G' \otimes_{A'} A \to G$. Indeed, injectivity follows immediately from flatness and for the converse, suppose that this map is injective and consider the exact sequence of A'-modules $0 \to \mathfrak{m} \to A \to \mathbb{C} \to 0$, where $\mathfrak{m}$ is the maximal ideal of A. We then have $F' \otimes_{A'} A = F$ and $\mathrm{Tor}_1^{A'}(F', A) = 0$. There is an exact sequence

$$\mathrm{Tor}_1^{A'}(F', A) \longrightarrow \mathrm{Tor}_1^{A'}(F', \mathbb{C}) \longrightarrow F' \otimes_{A'} \mathfrak{m} \xrightarrow{\ j\ } F' \otimes_{A'} A$$

and the map j can be identified with the map $F \otimes_A \mathfrak{m} \to F$ which is injective by the flatness assumption on F. Hence, $\mathrm{Tor}_1^{A'}(F', \mathbb{C}) = 0$ and the quotient F' is S'-flat by Proposition 4.1.2.

So we consider an S'-flat quotient $F' = \mathbf{E}'/G'$ which extends F. Then the sub-sheaf G' is S'-flat and by applying the functor $\otimes_{A'}$ to the exact sequence $0 \to I \to A' \to A \to 0$ we obtain the commuting diagram of short exact sequences,

$$\begin{array}{ccccccccc} & & 0 & & 0 & & 0 & & \\ & & \downarrow & & \downarrow & & \downarrow & & \\ 0 & \longrightarrow & G\otimes_A I & \longrightarrow & G' & \longrightarrow & G & \longrightarrow & 0 \\ & & \downarrow & & \downarrow & & \downarrow & & \\ 0 & \longrightarrow & \mathbf{E}\otimes_A I & \longrightarrow & \mathbf{E}' & \longrightarrow & \mathbf{E} & \longrightarrow & 0 \\ & & \downarrow & & \downarrow & & \downarrow & & \\ 0 & \longrightarrow & F\otimes_A I & \longrightarrow & F' & \longrightarrow & F & \longrightarrow & 0 \\ & & \downarrow & & \downarrow & & \downarrow & & \\ & & 0 & & 0 & & 0 & & \end{array}$$

from which we obtain $G' \cap (\mathbf{E} \otimes_A I) = G \otimes_A I$. In other words, the quotient $G'/G \otimes_A I$ splits the extension of G by $F \otimes_A I$ associated to the sheaf extension $0 \to \mathbf{E} \otimes_A I \to \mathbf{E}' \to \mathbf{E} \to 0$ over $X_{S'}$, via the inclusion $G \hookrightarrow \mathbf{E}$ and the projection $E \otimes_A I \to F \otimes_A I$:

$$\begin{array}{ccccccccc} & & & & & \swarrow & G & & \\ & & & & & & \downarrow & & \\ 0 & \longrightarrow & F\otimes_A I & \longrightarrow & \mathbf{E}'/G\otimes_A I & \longrightarrow & \mathbf{E} & \longrightarrow & 0 \end{array}$$

Conversely, by taking its inverse image, such a splitting gives rise to a sub-sheaf $G' \subset \mathbf{E}'$ over $X_{S'}$ such that $G' \cap (\mathbf{E} \otimes_A I) = G \otimes_A I$ and projects to G in $\mathbf{E}$. Then the canonical map $G' \otimes_{A'} A \to \mathbf{E}$ injects and has image G. The quotient $F' = \mathbf{E}'/G'$ is an S'-flat coherent sheaf which extends F.

Note that this extension is *a priori* over $X_{S'}$, but, in fact, it comes from X_S. Indeed, in computing $\mathrm{Ext}^1_{A'}(G, F \otimes I)$ from a locally-free resolution $R_\bullet \to G$ of G over $X_{S'}$, where R_i is a direct sum of invertible sheaves $\mathcal{O}_{X'_S}(-m_i)$ for sufficiently large integers m_i, shows that we have a natural exact sequence

$$0 \longrightarrow \mathrm{Ext}^1_A(G, F\otimes I) \longrightarrow \mathrm{Ext}^1_{A'}(G, F\otimes I) \to \mathrm{Hom}_A(\mathcal{T}or_1^{A'}(G, A), F\otimes I))$$

and this sequence fits into a commuting diagram

$$\begin{array}{ccccc} & & \mathrm{Ext}^1_{A'}(E, E\otimes_A I) & \to & \mathrm{Hom}_A(\mathcal{T}or_1^{A'}(E, A), E\otimes I) \\ & & \downarrow & & \downarrow \\ 0 \longrightarrow \mathrm{Ext}^1_A(G, F\otimes I) & \longrightarrow & \mathrm{Ext}^1_{A'}(G, F\otimes I) & \longrightarrow & \mathrm{Hom}_A(\mathcal{T}or_1^{A'}(G, A), F\otimes I) \end{array}$$

in which the second horizontal map takes an extension to the associated boundary map in the derived exact sequence of the functor $-\otimes_{A'} A$.

Let $e \in \mathrm{Ext}^1_{A'}(\mathbf{E}, \mathbf{E} \otimes_A I)$ be given by the canonical extension. From the fact that $\mathbf{E}'$ is A'-flat, the isomorphism of functors

$$(- \otimes_{A'} \mathbf{E}') \otimes_{A'} A \simeq - \otimes_A \mathbf{E}$$

shows that the boundary homomorphism associated to e is also, up to sign, the boundary homomorphism associated to the exact sequence $0 \to I \to A' \to A \to 0$ in the derived exact sequence of the functor $- \otimes_{A'} \mathbf{E}$. The commuting diagram

$$\begin{array}{ccc} \mathcal{T}or_1^{A'}(G, A) & \longrightarrow & G \otimes_A I \\ \downarrow & & \downarrow \\ \mathcal{T}or_1^{A'}(\mathbf{E}, A) & \longrightarrow & \mathbf{E} \otimes I \\ & & \downarrow \\ & & F \otimes_A I \end{array}$$

then shows that the image of e in $\mathrm{Hom}_A(\mathcal{T}or_1^{A'}(G, A), F \otimes I)$ vanishes and this proves our claim.

It follows that if $\mathrm{Ext}^1_A(G, F \otimes_A I) = 0$ then such an extension always exists. Moreover, the set of such extensions carries an affine space structure with tangent vector space $\mathrm{Hom}_A(F, F \otimes_A I)$.

We apply this to the case where $S = \bullet$ is a smooth point and where $S' = D$ is associated to the algebra of dual numbers. The composite $\bullet \to D \to \bullet$ is obviously the identity and the ideal I is isomorphic to $\mathbb{C}$. It follows that each quotient F of E can be canonically extended to D: it suffices to choose F' to be the inverse image of F under the projection $D \to \bullet$. Another way to see this is to note that, on X, the exact sequence of sheaves $0 \to \mathbf{E} \otimes_A I \to \mathbf{E}' \to \mathbf{E} \to 0$ is canonically split. It follows that we obtain a canonical isomorphism

$$T_q \,\mathbf{Hilb}^{r,d}(E) = \mathrm{Hom}(G, F),$$

where q is the point of the Hilbert scheme given by the quotient $F = E/G$. The two sides of this equality are vector spaces; we shall see below that this isomorphism is actually linear.

To show that the Hilbert scheme is smooth at the point q given by the quotient $F_0 = E/G_0$, it suffices to show that if $F = \mathbf{E}/G$ is an S-flat extension of F_0 over a fat point S then we have $\mathrm{Ext}^1(G, F \otimes I) = 0$. We shall prove that this is the case under the second hypothesis, that is, if $\mathrm{Ext}^1(G_0, F_0) = 0$. We can obviously assume, by proceeding case by case, that $\mathfrak{m}I = 0$ which implies that I is an A-module which is associated to a point. Hence, $F \otimes_A I \simeq F_0 \otimes_{\mathbb{C}} I$.

On the other hand, the sub-sheaf G is locally-free over X_S. Indeed, $G \otimes_A \mathbb{C} = G_0$ is locally-free over X. If x is a point of X then an isomorphism $\mathcal{O}^k_{X_S,x} \otimes_A \mathbb{C} \simeq G_x$ lifts to a map $\mathcal{O}^k_{X_S,x} \to G_x$ which surjects by Nakayama's Lemma. Since G is flat over S, the kernel N of this morphism also satisfies the condition $N_x \otimes \mathbb{C} = 0$ and so $N_x = 0$ again by Nakayama's Lemma.

We then have

$$\begin{aligned}\mathrm{Ext}^1(G, F_0 \otimes_{\mathbb{C}} I) &= H^1(X_S, \mathcal{H}om_A(G, F_0 \otimes_{\mathbb{C}} I)) \\ &= H^1(X, \mathcal{H}om(G_0, F_0 \otimes_{\mathbb{C}} I)) \\ &= \mathrm{Ext}^1(G_0, F_0) \otimes_{\mathbb{C}} I.\end{aligned}$$

Hence, the hypothesis $\mathrm{Ext}^1(G_0, F_0) = 0$ implies that the Hilbert scheme is smooth at the point q given by F_0. □

Lemma 8.2.3 *Let $q \in \mathbf{Hilb}^{r,d}(E)$ be a point of the Hilbert scheme corresponding to a coherent quotient sheaf $F = E/G$. The isomorphism $T_q\,\mathbf{Hilb}^{r,d}(E) \to \mathrm{Hom}(E, F)$ constructed above is linear.*

Proof Let $\mathbf{H} = \mathbf{Hilb}^{r,d}(E)$ be the Hilbert scheme, $\mathbf{F} = \mathbf{E_H}/\mathbf{G}$ be the universal quotient sheaf on $X_{\mathbf{H}}$. Let S be the infinitesimal neighbourhood of order 1 in $\mathbf{H}$ of the point q. The point q is defined in S by an ideal I of square zero which is isomorphic to the cotangent space $T^*_q\mathbf{H}$ and the projection $S \to \{q\}$ gives us a "constant" quotient sheaf by pullback. On the other hand, we have at our disposal the restriction of the universal quotient $\mathbf{F}|_{X_S}$ and the construction given in the proof of Theorem 8.2.1 gives us a universal element

$$\omega \in \mathrm{Hom}(G, F \otimes I) = \mathrm{Hom}(G, F \otimes T^*_q\mathbf{H}).$$

The tangent space $T_q\mathbf{H}$ can be identified with the space of morphisms $\mathrm{Mor}_q(D, \mathbf{H})$. These morphisms factor uniquely through the infinitesimal neighbourhood S. We now note by functoriality that for each tangent vector $t \in T_q\mathbf{H}$, the element $\omega(t) \in \mathrm{Hom}(G, F)$ is just the image of t under the isomorphism of Theorem 8.2.1. This gives us the linearity. □

The action of the group $\mathrm{Aut}(E)$

The algebraic group $\mathrm{Aut}(E)$ of automorphisms of E acts on the Hilbert scheme $\mathbf{Hilb}^{r,d}(E)$: indeed, for every algebraic variety S, the group of morphisms $\mathrm{Mor}(S, \mathrm{Aut}(E))$ can be identified with the automorphism group of the locally-free sheaf $\mathbf{E} = E_S$ and this group acts on the set $\mathbf{Hilb}^{r,d}(E)$ of S-flat quotient sheaves $F = E_S/G$ which give rise to quotient sheaves of rank r and degree d over each point $s \in S$ (cf. chapter 4). Indeed, if $g \in \mathrm{Mor}(S, \mathrm{Aut}(E))$ and $F = E_S/G \in \mathbf{Hilb}^{r,d}(E; S)$ then we can consider the image sub-sheaf $g(G)$: the quotient $g(F) = E_S/g(G)$ is still S-flat and belongs to $\mathbf{Hilb}^{r,d}(E; S)$. Hence, we have obtained a morphism which is functorial in S

$$\mathrm{Mor}(S, \mathrm{Aut}(E)) \times \mathbf{Hilb}^{r,d}(E; S) \longrightarrow \mathbf{Hilb}^{r,d}(E; S)$$

from which we get the map of algebraic varieties $\mathrm{Aut}(E) \times \mathbf{Hilb}^{r,d}(E) \to \mathbf{Hilb}^{r,d}(E)$ and this is clearly a group action. We have used this action several times in the previous chapter when E was the trivial bundle.

Proposition 8.2.4 *Let $q \in \mathbf{Hilb}^{r,d}(E)$ be a point corresponding to a coherent quotient sheaf $F = E/G$. Let $p : E \to F$ be the canonical projection. The tangent linear map at the identity element of the orbit map $\mathrm{Aut}(E) \to \mathbf{Hilb}^{r,d}(E)$ is the linear map*

$$\mathrm{Lie}(\mathrm{Aut}(E)) = \mathrm{End}(E) \longrightarrow \mathrm{Hom}(G, F)$$

given by $v \mapsto p \circ v|_G$.

Proof The Lie algebra $\mathrm{End}(E)$ can be identified with $\mathrm{Mor}(D, \mathrm{Aut}(E)) = \mathrm{Aut}(E_D)$ by associating the map $u = \mathrm{id}_E + \epsilon v$ to $v \in \mathrm{End}(E)$. So we obtain an automorphism of the exact sequence $0 \to E \otimes I \to E_D \to E \to 0$, where $I = \mathbb{C}\epsilon$ is the maximal ideal of the ring of dual numbers. The image of v under the tangent map to the orbit of q

$$\mathrm{Mor}(D, \mathrm{Aut}(E)) \longrightarrow \mathrm{Mor}(D, \mathbf{Hilb}^{r,d}(E))$$

corresponds to the image quotient module E_D/G' of the canonical section $G \mapsto E_D/G \otimes I$ under the automorphism u. It is clear that the associated morphism $G \to F$ is just $p \circ v|_G$. This proves the proposition. □

8.3 Smoothness of $\mathbf{M}^s(r,d)$

Let H be a vector space and $E = H \otimes \mathcal{O}_X$. Suppose that the integer d is at least $r(2g-1)$ and consider the open subset Ω of $\mathbf{Hilb}^{r,d}(E)$ of points q representing semi-stable quotient sheaves $F = E/G$ such that the canonical map $H \to H^0(F)$ is an isomorphism: this open set is smooth by Corollary 8.2.2. We also have

Lemma 8.3.1 *Let $q \in \Omega$ and $F = E/G$ be the corresponding quotient sheaf. The stabilizer of q in $\mathrm{GL}(H)$ is isomorphic to the automorphism group of the sheaf F.*

Proof An element g of $\mathrm{GL}(H)$ leaves q fixed if $g(G) = G$. Such an element therefore induces an automorphism $\bar{g}$ of F. Conversely, an automorphism $\varphi : F \to F$ lifts to an automorphism g of H and from the isomorphism $H \simeq H^0(F)$ the commuting diagram

$$\begin{array}{ccc} H \otimes \mathcal{O}_X & \longrightarrow & F \\ {\scriptstyle g}\downarrow & & \downarrow{\scriptstyle \varphi} \\ H \otimes \mathcal{O}_X & \longrightarrow & F \end{array}$$

shows that $\bar{g} = \varphi$. Hence, the group homomorphism $g \mapsto \bar{g}$ surjects. The injectivity follows from the commuting diagram

$$\begin{array}{ccc} H & \longrightarrow & H^0(F) \\ {\scriptstyle g}\downarrow & & \downarrow{\scriptstyle \bar{g}} \\ H & \longrightarrow & H^0(F) \end{array}$$

This completes the proof. □

Theorem 8.3.2 *The variety $\mathbf{M}^s(r,d)$, when it is not empty, is smooth of dimension $r^2(g-1)+1$. At a point $a \in \mathbf{M}^s(r,d)$ given by a stable bundle F we have a canonical isomorphism*

$$T_a\mathbf{M}^s(r,d) \simeq \mathrm{Ext}^1(F,F).$$

It follows from Lemma 8.3.1 that the $\mathrm{GL}(H)$ action factors through an action of $\mathrm{PGL}(H)$ and this action has no fixed points in the open set Ω^s corresponding to the points of Ω which are stable.

Lemma 8.3.3 *The action of $G = \mathrm{PGL}(H)$ on the open set Ω^s is free.*

Recall that this means that the graph map $G \times \Omega^s \to \Omega^s \times \Omega^s$ is an isomorphism on a closed sub-variety R of $\Omega^s \times \Omega^s$.

Proof The fibres of the graph map are reduced to points by Lemma 8.3.1. For a neat proof of the result we could appeal to a general result, due to Mumford, which reduces the statement to studying the action of one parameter sub-groups [42, Proposition 2.4 and Corollary 2.5]. The proof which we propose to give here is more direct. Let $\mathbf{F}$ be the universal quotient bundle on $\Omega^s \times X$. The graph R of the equivalence relation is the set of pairs $(q,q') \in \Omega^s \times \Omega^s$ such that $\mathbf{F}(q) \simeq \mathbf{F}(q')$, which we can rewrite, because of the stability of the bundles $\mathbf{F}(q)$, as

$$\operatorname{Hom}(\mathbf{F}(q), \mathbf{F}(q')) \neq 0.$$

From the Semi-continuity Theorem, R is a closed subset of $\Omega^s \times \Omega^s$. It suffices to show that the graph map $G \times X \to R$ admits local sections. Consider the two projection maps $\mathrm{pr}_i : R \to \Omega^s$ and the pullback bundles $\mathbf{F}_i = (\mathrm{pr}_i \times \mathrm{id}_X)^*(\mathbf{F})$ over $R \times X$. On the other hand, let $p : R \times X \to R$ be the projection to the first factor. Since the variety R is reduced by definition, the direct image $\mathcal{L} = p_*(\mathcal{H}om(\mathbf{F}_1, \mathbf{F}_2))$ is locally-free of rank one, the fibre of $\mathcal{L}$ over the point (q,q') can be identified with $\operatorname{Hom}(\mathbf{F}(q), \mathbf{F}(q'))$ and the evaluation map

$$p_*(\mathcal{H}om(\mathbf{F}_1, \mathbf{F}_2)) \otimes \mathbb{C} \longrightarrow \operatorname{Hom}(\mathbf{F}(q), \mathbf{F}(q'))$$

is an isomorphism (cf. Corollaries 4.2.9 and 4.2.11). Hence, the choice of a local trivialization of $\mathcal{L}$ over an open set $U \subset R$ gives an isomorphism

$$\mathbf{F}_1|_{U\times X} \simeq \mathbf{F}_2|_{U\times X}.$$

This isomorphism lifts to an automorphism of the trivial bundle $H \times \mathcal{O}_{U\times X}$ which gives rise to a map $g : (q,q') \mapsto g_{q,q'}$ of U to $\mathrm{GL}(H)$ such that $g_{q,q'}(q) = q'$. Thus, we have a local section $(q,q') \mapsto (q, g_{q,q'})$ and it follows that the graph map is an isomorphism onto its image. □

Proposition 8.3.4 *Let G be a linearly reductive algebraic group acting freely on a smooth algebraic variety Y. Suppose that Y has a geometric quotient $\varphi : Y \to Z$. Then*

- *(i) the algebraic variety Z is smooth,*
- *(ii) the map $Y \to Z$ is a submersion and its fibres are isomorphic to G, and*
- *(iii) the diagram*

$$\begin{array}{ccc} Y \times G & \xrightarrow{\sigma} & Y \\ {\scriptstyle \mathrm{pr}_1}\downarrow & & \downarrow{\scriptstyle \varphi} \\ Y & \xrightarrow{\varphi} & Z \end{array}$$

where σ is the map describing the action of G is cartesian.

The last part means that the embedding of $Y \times G$ into the fibre product $Y \times_Z Y$ is an isomorphism. Since $Y \times_Z Y$ is smooth by property (ii), this part follows immediately from the fact that the action is free.

Corollary 8.3.5 *Under the same hypotheses, if $a \in Y$ has image $b = \varphi(a)$ in Z then the Zariski tangent space $T_b Z$ at the point b is canonically isomorphic to the normal space N_a of the orbit $\mathrm{Orb}^G a$ of the point a; this isomorphism is compatible with the G action.*

Corollary 8.3.6 *Under the same hypotheses, there exist étale maps $\rho : V \to Z$ whose images cover Z and G-equivariant maps $\tau : V \times g \to Y$ such that the diagrams*

$$\begin{array}{ccc} V \times G & \xrightarrow{\tau} & Y \\ {\scriptstyle \mathrm{pr}_1}\downarrow & & \downarrow{\scriptstyle \varphi} \\ V & \xrightarrow{\rho} & Z \end{array}$$

are cartesian.

In other words, should $\varphi : Y \to Z$ not be a locally trivial principal G-bundle in the Zariski topology then it will become one when we pass to étale covers; we also say that φ is *locally isotrivial.* This result follows easily from part (iii) above. It suffices to choose a germ of a smooth sub-variety of the same dimension as Z in the neighbourhood of a point $a \in Y$ which passes through a and is transverse to the orbit through a; perhaps after shrinking V, we can assume that $\varphi|_V$ is étale. We come now to the proof of Proposition 8.3.4. We use the following fact:

Lemma 8.3.7 *Suppose that the group G is linearly reductive. Let $\varphi : Y \to Z$ be a good quotient and let $Z' \subset Z$ be an algebraic sub-variety. Then the map $\varphi' : Y' = Z' \times_Z Y \to Z'$ is a good quotient.*

Proof The question is local on Z and so we are reduced to the case where Z is affine. In this case, Y is affine and we have $Y = \operatorname{Spec} A$, where A is an algebra of finite type on which G acts. Then $Z = \operatorname{Spec} A^G$. The sub-variety Z' is given by an ideal J of A^G and so we have $Z' = \operatorname{Spec} A^G/J$. The sub-variety Y' of Y is given by the ideal JA and, by Lemma 6.3.5, $JA \cap A^G = J$. It follows from Proposition 6.2.6 that $(A/JA)^G = A^G/J$. The map $Y' \to Z'$ is induced by the inclusion $(A/JA)^G \hookrightarrow A/JA$ and it follows from Theorem 6.3.1 that this is a good quotient. □

Proof of Proposition 8.3.4 The question is local on Z so we assume that Z is an affine variety. The same is true of Y which we can assume to be smooth, irreducible and of dimension n. Let g be the dimension of G. Since the action is free, all of the fibres of φ have G as their underlying reduced variety and they all have the same dimension g. Thus, Z is an irreducible variety of dimension $n-g$. We shall first show that the fibre Y_b over a point $b \in Y$ is actually smooth. Consider a point a such that $\varphi(a) = b$ and a locally closed smooth sub-variety Σ meeting $\mathrm{Orb}^G a$ only at the point a and given by an ideal $(z_1, \ldots, z_g)$, where z_i are regular functions defined in a neighbourhood of a in Y giving a system of parameters for the orbit $\mathrm{Orb}^G a$ of a. This sub-variety has codimension $g = \dim G$ and is transverse to the orbit. Consider the intersection $\Sigma_b = \Sigma \cap Y_b$. Then Σ_b is a fat point.

Lemma 8.3.8 *The map*

$$\Sigma_b \times G \xrightarrow{\psi} Y_b \tag{8.1}$$

is an isomorphism.

Since this map is G-equivariant, Σ_b is a good quotient of Y_b due to the preceding Lemma 8.3.7 and the uniqueness of good quotients. This implies that Σ_b is reduced at the smooth point a and so the map $G \to Y_b$ induced by the action of the group is an isomorphism. We can then use Lemma 8.3.7 to observe that the infinitesimal neighbourhood of order 1 of the orbit $\mathrm{Orb}^G a$ in Y has a good quotient which is just the infinitesimal neighbourhood of order 1 of b. It follows that if $\mathfrak{m}_b$ denotes the maximal ideal of b in Z then the canonical map

$$\mathfrak{m}_b/\mathfrak{m}_b^2 \longrightarrow (\mathfrak{m}_b\mathcal{O}(Y)/\mathfrak{m}_b^2\mathcal{O}(Y))^G$$

is an isomorphism. The right hand side is just the space of G-invariant sections of the conormal bundle N^* of the orbit $\mathrm{Orb}^G a$ in Y. Such a section is determined by its value at a. Therefore, its vector space of sections has dimension $n-g$. Hence, the tangential dimension of Z is $n-g$ at every point and so the variety Z is smooth. Moreover, the map induced by the linear cotangent map

$$T_b^*(Z) \longrightarrow N_a^* \hookrightarrow T_a^*(Y)$$

injects which means that the map φ is a submersion. This verifies parts (i) and (ii) of the proposition and we have already seen in the preamble that (iii) is immediate. □

Proof of Lemma 8.3.8 We use the notion of infinitesimal neighbourhood of order n of the orbit $\mathrm{Orb}^G a$ in Y: if I is the ideal defining the orbit in Y then this is just the algebraic variety $\mathrm{Spec}\, \mathcal{O}(Y)/I^{n+1}$. Similarly, we can define the infinitesimal neighbourhood of order n of $\{a\} \times G$ in $\Sigma \times G$. The map

$$\sigma : \Sigma \times G \to Y$$

induced by the group action is étale along $\{a\} \times G$ and is an embedding of this sub-variety so this map σ induces an isomorphism of infinitesimal neighbourhoods of order n (the proof is identical to that of Lemma 8.1.8). The two sides of (8.1) are obviously contained in the infinitesimal neighbourhoods of order n above for n sufficiently large, and so ψ is an embedding. Moreover, if Σ'_b denotes the fibre of the composite

$$\begin{array}{ccc} (\Sigma_b \times G, (a, e)) & \xrightarrow{\quad \psi \quad} & (Y_b, a) \\ & \underset{\zeta \circ \psi}{\searrow} & \downarrow \zeta \\ & & (\mathbb{C}^g, 0) \end{array}$$

where ζ is the map $(z_1, \ldots, z_g)$, then this implies that the induced map $\psi : (\Sigma'_b, (a, e)) \to (\Sigma_b, a)$ is an embedding. Since the composite $(\Sigma_b, a) \to (\Sigma'_b, (a, e)) \to (\Sigma_b, a)$ is the identity, this map is forced to be an isomorphism and Σ'_b coincides with Σ_b. We further remark that the map $\zeta \circ \psi$ is flat at (a, e): this follows from the fact that $(\psi * (z_i))_{i=1,\ldots,g}$ is a regular sequence in the Cohen–Macaulay local ring $\mathcal{O}_{\Sigma_b \times G, (a,e)}$ (cf. Hartshorne [31], chapter II, theorem 8.21 and chapter III, lemma 10.3).

To see that ψ is actually an isomorphism, note that it is clearly a homeomorphism and so a finite morphism. The sheaf $\psi_*(\mathcal{O}_{\Sigma_b \times G})$ is then a coherent algebraic sheaf on Y_b. We need to see that the map

$$\mathcal{O}_{Y_b} \longrightarrow \psi_*(\mathcal{O}_{\Sigma_b \times G})$$

of coherent sheaves on Y_b, which we know to be an epimorphism in a neighbourhood of the point a, is injective. We have seen that it induces an isomorphism in the fibre over 0; this implies that if $\mathcal{N}$ is its kernel then we have $\mathcal{N} \otimes_{\mathcal{O}_{\mathbb{C}^p,0}} \mathbb{C} = 0$ due to the flatness of $\Sigma_b \times G$ over $\mathbb{C}^p$ in a neighbourhood of a, and so $\mathcal{N} = 0$ in a neighbourhood of a by Nakayama's Lemma. Hence, the map ψ is an isomorphism in a neighbourhood of a and, since it is equivariant, it is an isomorphism. This completes the proof of Lemma 8.3.8. □

Proof of Theorem 8.3.2 It follows from Proposition 8.3.4 above that the quotient $\mathbf{M}^s(r,d)$ of Ω^s is smooth and the tangent space at a point a, which is the image of a point q represented by the quotient sheaf $F = E/G$, is isomorphic to the vector space normal to the orbit of the point q and so, by Proposition 8.2.4, it is the cokernel of the linear map

$$\begin{array}{ccc} \mathrm{End}(H) & \longrightarrow & \mathrm{Hom}(G,F): \\ v & \longmapsto & p \circ g|_G, \end{array}$$

where $p : E \to F$ denotes the canonical projection. Consider the exact sequence

$$0 \longrightarrow G \longrightarrow H \otimes \mathcal{O}_X \longrightarrow F \longrightarrow 0$$

and apply the functor $\mathrm{Hom}(-,F)$ to it. From the fact that $\mathrm{Ext}^1(H \otimes \mathcal{O}_X, F) = 0$ because the degree d is at least $r(2g-1)$, we obtain the exact sequence

$$\mathrm{Hom}(H \otimes \mathcal{O}_X, F) \xrightarrow{\text{restriction}} \mathrm{Hom}(G,F) \longrightarrow \mathrm{Ext}^1(F,F) \longrightarrow 0$$

and the projection p induces an isomorphism $\mathrm{End}(H) \simeq \mathrm{Hom}(H \otimes \mathcal{O}_X, F)$. It follows that the boundary map induces an isomorphism between the normal space to the orbit at q and $\mathrm{Ext}^1(F,F)$. Moreover, if $q' = g(q)$ with $g \in \mathrm{PGL}(H)$ is another point in the orbit representing a quotient $F = E/G'$ then the commuting diagram of exact sequences

$$\begin{array}{ccccccccc} 0 & \longrightarrow & G & \longrightarrow & H \otimes \mathcal{O}_X & \longrightarrow & F & \longrightarrow & 0 \\ & & {\scriptstyle g}\downarrow & & {\scriptstyle g}\downarrow & & {\scriptstyle \mathrm{id}_F}\downarrow & & \\ 0 & \longrightarrow & G' & \longrightarrow & H \otimes \mathcal{O}_x & \longrightarrow & F & \longrightarrow & 0 \end{array}$$

gives rise to the commuting diagram

$$\begin{array}{ccc} \mathrm{Hom}(G',F) & \longrightarrow & \mathrm{Ext}^1(F,F) \\ {\scriptstyle u \mapsto u \circ g}\downarrow & & \| \\ \mathrm{Hom}(G,f) & \longrightarrow & \mathrm{Ext}^1(F,F) \end{array}$$

in which the horizontal maps are the boundary maps. At the point a of $\mathbf{M}^s(r,d)$ defined by F, the identification $T_a\mathbf{M}^s(r,d) \simeq \mathrm{Ext}^1(F,F)$ we have found above is therefore canonical. It remains to compute its dimension. The endomorphism bundle $\mathcal{H}om(F,F)$ has rank r^2 and degree 0 and so, by the Riemann-Roch formula, $\chi(\mathcal{H}om(F,F)) = r^2(1-g)$. On the other hand, the bundle F is assumed to be stable, consequently it has no endomorphisms other than homotheties. Hence, $\dim \mathrm{Ext}^1(F,F) = r^2(g-1)+1$. □

8.4 The universal family

Definition 8.4.1 *Let S be an algebraic variety and $f : S \to \mathbf{M}(r,d)$ a morphism. By* universal bundle *(or* Poincaré bundle*) on $S \times X$ we mean a vector bundle $\mathbf{F} \in \mathbf{M}(r,d)(S)$ such that the induced morphism $S \to \mathbf{M}(r,d)$ is f.*

Theorem 8.4.2 *If r and d are coprime then there exists a universal bundle on $\mathbf{M}(r,d) \times X$.*

In other words, the functorial morphism

$$\mathbf{M}(r,d)(S) \longrightarrow \mathrm{Mor}(S, \mathbf{M}(r,d))$$

surjects for all algebraic varieties S. Notice that such universal bundles are not unique because if $\mathbf{U}$ is a universal bundle and L is an invertible sheaf on $\mathbf{M}(r,d)$ then the bundle $\mathbf{U} \otimes \mathrm{pr}_1^* L$ is still a universal bundle. The proof depends on a lemma telling us when a vector bundle descends to a quotient.

Definition 8.4.3 *Let G be an algebraic group acting on an algebraic variety Y. An* algebraic vector G-bundle *$F \to Y$ is a vector bundle over Y equipped with a G action which is linear in each fibre and such that the diagram*

$$\begin{array}{ccc} G \times F & \longrightarrow & F \\ \downarrow & & \downarrow \\ G \times Y & \longrightarrow & Y \end{array}$$

commutes.

Lemma 8.4.4 *Let G be a linearly reductive group acting freely on an algebraic variety Y. Suppose that there exists a geometric quotient $\varphi : Y \to Y'$. Let $F \to Y$ be an algebraic vector G-bundle over Y. Then there exists a geometric quotient $F' = F/G$ which is a vector bundle over Y'.*

Proof The question is local on Z. If O is an orbit then the vector G-bundle $F|_O$ is obviously trivial. Such a G-trivialization extends to an affine open neighbourhood U of O to give a trivialization

$$\eta : U \times \mathbb{C}^r \simeq F|_U.$$

We can assume that U is invariant and, after applying the Reynold's operator and, if necessary, shrinking the open set U, we can also assume that the trivialization η is G-invariant. We are reduced to showing that that there exists a quotient for $U \times \mathbb{C}^r$ with the product action where the group G acts trivially on the factor $\mathbb{C}^r$. But the product $U/G \times \mathbb{C}^r$ is obviously a geometric quotient which establishes the first assertion. If we change the open set U then the change in trivialization is evidently given by a G-invariant map with values in $\mathrm{GL}(r, \mathbb{C})$ which gives rise to a vector bundle structure. □

The vector space of sections of F' over an open set U' is clearly the space $\Gamma(\varphi^{-1}(U'), F)^G$ of invariants: this can be checked locally. In particular, this proves that the sheaf of modules on Y' given locally by the presheaf $U' \mapsto \Gamma(\varphi^{-1}(U'), F)^G$ is locally-free.

Exercise (Kempf's Lemma). Let G be a linearly reductive group acting on an algebraic variety Y. Suppose that there exists a good quotient $\varphi : Y \to Y'$. Let $F \to Y$ be an algebraic vector G-bundle over Y. If the stabilizer of a point in a closed orbit acts trivially on F then there exists a geometric quotient $F' = F/G$ and this is a vector bundle over Y'.

Proof of Theorem 8.4.2 We can certainly assume that $d > 2rg$. Note also that, under the hypothesis that r and d are coprime, the semi-stable bundles of rank r and degree d are stable. Consider the universal quotient bundle $\mathbf{F}$ on the open subset $\Omega \times X$ of $\mathbf{Hilb}^{r,d}(H \otimes \mathcal{O}_X) \times X$. The group $\mathrm{SL}(H)$ acts on $\mathbf{F}$ and so we obtain a vector $\mathrm{SL}(H)$-bundle. It is certainly not possible to take the quotient of this vector bundle because the stabilizer of a point (the group of homotheties given by the χth roots of unity) does not act trivially on $\mathbf{F}$. We must find an invertible vector $\mathrm{SL}(H)$-bundle $\mathcal{L}$ over Ω so that the action of the stabilizer on $\mathbf{F}' = \mathbf{F} \otimes \mathrm{pr}_1^* \mathcal{L}$ is trivial. Indeed, the quotient vector bundle $\mathbf{F}' = \mathbf{F}/\mathrm{SL}(H)$ then makes sense over $\mathbf{M}(r,d) \times X$ and defines a universal bundle.

The integers r and $\chi = d + r(1-g)$ are coprime. We choose two integers u and v such that $ur + v\chi = 1$ and fix a point $a \in X$. Consider the bundles over Ω of respective ranks r and χ given by

$$\mathbf{F}_1 = \mathbf{F}(a) \text{ and } \mathbf{F}_2 = \mathrm{pr}_{1*}(\mathbf{F}).$$

The invertible bundle defined by $\mathcal{L} = (\det \mathbf{F}_1)^{\otimes u} \otimes (\det \mathbf{F}_2)^{\otimes v}$ is a bundle with an $\mathrm{SL}(H)$ action and the centre of $\mathrm{SL}(H)$ acts on this bundle and can be identified as the group of homotheties of $\mathcal{L}$ with ratios the χth roots of unity. It follows that the action of the stabilizer on the bundle $\mathbf{F}' = \mathbf{F} \otimes \mathrm{pr}_1^*(\mathcal{L}^*)$ is trivial and this yields the required universal bundle. □

8.5 Irreducibility

We are now going to show that the moduli space is irreducible when the rank is 1. The general case can be reduced to this one by expressing any semi-stable bundle of rank r as an extension of a rank 1 bundle by the trivial bundle of rank $r-1$.

The case of invertible bundles

The moduli space of isomorphism classes of invertible bundles of degree d is a smooth projective variety of dimension g, is denoted $\mathrm{Pic}^d(X)$ and admits a Poincaré bundle over $\mathrm{Pic}^d(X) \times X$. Actually, the tensor product gives rise to a group structure on $J(X) = \mathrm{Pic}^0(X)$: this is an abelian variety called the *Jacobian* of X. All of the varieties $\mathrm{Pic}^d(X)$ are isomorphic to $\mathrm{Pic}^0(X)$.

Proposition 8.5.1 *The variety $J(X)$ is irreducible.*

Proof For $d \geq g$, consider the variety $\mathrm{Div}^d(X)$ of effective divisors of degree d. This is a smooth variety of dimension d and is obviously irreducible. Moreover, we have a canonical map

$$\mathrm{Div}^d(X) \longrightarrow \mathrm{Pic}^d(X)$$

which associates the invertible sheaf L_D defined in chapter 2 to a divisor D. This is a morphism of algebraic varieties. Indeed, there exists a universal divisor on the product $\mathrm{Div}^d(X) \times X$, to which we can associate an invertible bundle of relative degree d. The coarse moduli property of $\mathrm{Pic}^d(X)$ then gives us the required map. Since $d \geq g$, each invertible bundle of degree d has at least one section by the Riemann-Roch Theorem. Hence, the above map surjects. The irreducibility of $\mathrm{Pic}^d(X)$ now follows and so too for $J(X)$. □

The arbitrary r case

Theorem 8.5.2 *The moduli variety $\mathbf{M}(r,d)$ is irreducible.*

Proof Consider the extensions

$$0 \longrightarrow \mathcal{O}_X^{r-1} \longrightarrow F \longrightarrow L \longrightarrow 0, \tag{8.2}$$

where L is an invertible bundle of degree d and we assume that $d \geq 2rg$. These extensions are classified, for a fixed L, by $\mathrm{Ext}^1(L, \mathcal{O}_X^{r-1}) = H^1(X, L^*)^{r-1}$ a vector space of dimension $(r-1)(d+g-1)$. As L varies over $\mathrm{Pic}^d(X)$ these vector spaces form a variety which can be identified with the total space of the vector bundle $\mathcal{V}$ over $\mathrm{Pic}^d(X)$ given by

$$\mathcal{V} = R^1 \mathrm{pr}_{1*}(\mathcal{L}^*)^{r-1},$$

where $\mathcal{L}$ is a Poincaré bundle on $\mathrm{Pic}^d(X) \times X$. This coherent sheaf is locally-free by Corollary 4.2.11. Moreover, there exists a universal extension on $\mathcal{V} \times X$: this follows from section 7.3. Consider the open set $\mathcal{V}^{ss}$ of extensions for which F is semi-stable. The coarse moduli property then gives a map

$$f : \mathcal{V}^{ss} \longrightarrow \mathbf{M}(r,d)$$

associating to the above extension the S-equivalence class of the bundle F.

Lemma 8.5.3 *The map $f : \mathcal{V}^{ss} \to \mathbf{M}(r,d)$ associated to the universal extension above is surjective.*

The theorem now follows because the variety $\mathcal{V}$ is irreducible as $\mathrm{Pic}^d(X)$ is and so $\mathcal{V}^{ss}$ is irreducible. □

Proof of Lemma 8.5.3 Let F be a semi-stable bundle of degree d. From the choice of d, this bundle is generated by its sections. Then there exists a vector sub-space $\Gamma \subset H^0(F)$ of dimension $r-1$ such that the map induced by the evaluation map

$$\Gamma \otimes \mathcal{O}_X \longrightarrow F$$

has rank $r-1$ at all points. Indeed, if $a \in X$ is some point then the set of points $\Gamma \in \mathbf{Grass}(r-1, H^0(F))$ for which the linear map $\Gamma \to F(a)$ has rank less than $r-1$ is of codimension 2 in the Grassmannian (see the exercise below). The union of these closed sets as a varies is a closed set (Properness Theorem) of codimension at least 1. Hence, there exist points Γ which are not contained in any of these sub-spaces.

The cokernel L of such a map is then an invertible bundle. It follows that the vector bundle F fits into an extension of type (8.2), which establishes the surjectivity. □

Exercise. In the vector space $L(\mathbb{C}^{r-1}, \mathbb{C}^r)$ of linear maps $\mathbb{C}^{r-1} \to \mathbb{C}^r$, the set of maps which are not injective forms a closed irreducible subset of codimension exactly 2.

8.6 The existence problem

On a smooth rational projective algebraic curve X we know the structure of the algebraic vector bundles: Grothendieck's result (Lemma 4.4.1) shows that there are no stable bundles of rank bigger than 1. Moreover, there is a semi-stable bundle of rank r and degree d if and only if r divides d. Let r be an integer greater than or equal to 2.

Theorem 8.6.1 *If the curve X has genus g at least 2 then the open set $\mathbf{M}^s(r,d)$ is not empty for all pairs (r,d).*

It follows that the irreducible variety $\mathbf{M}(r,d)$ has dimension $r^2(g-1)+1$ since this holds for $\mathbf{M}^s(r,d)$. The situation is more delicate for the case of elliptic curves.

Theorem 8.6.2 *If X is an elliptic curve then the moduli space $\mathbf{M}(r,d)$ is non-empty. Moreover:*

(i) If r and d are coprime then $\mathbf{M}(r,d)$, the space of isomorphism classes of stable bundles, is a projective variety of dimension 1 isomorphic to X.

(ii) More generally, let ν be the greatest common divisor of r and d. Then the variety $\mathbf{M}(r,d)$ is a projective variety isomorphic to the variety $\mathrm{Div}^\nu(X)$ of effective divisors of degree ν on X.

The variety above is also the symmetric power $S^\nu X$. When r and d are coprime, Atiyah shows that the map

$$\begin{array}{ccc} \mathbf{M}(r,d) & \longrightarrow & \mathrm{Pic}^d(X) : \\ G & \longmapsto & \det F \end{array}$$

admits a section. By irreducibility, it follows that it is an isomorphism. The variety $\mathrm{Pic}^1(X)$ is canonically isomorphic to X and this leads to part (i). Part (ii) is an easy consequence of this (see the exercises below).

The relative Hilbert scheme

Let S be an affine algebraic variety and $E \to S \times X$ a vector bundle. For all morphisms $\alpha : S' \to S$ of algebraic varieties we set

$$E_{S'} = (\alpha \times \mathrm{id}_X)^*(E).$$

Consider the functor $S' \mapsto \mathbf{Hilb}^{r,d}(E/S, S')$ which associates to the algebraic variety S' the set of coherent quotient sheaves $\mathbf{F}'$ of $E_{S'}$ which are flat over S' and such that, for all $s \in S'$, the quotient $\mathbf{F}'(s)$ has rank r and degree d.

Proposition 8.6.3 *The functor $S' \mapsto \mathbf{Hilb}^{r,d}(E/S, S')$ is representable by an algebraic variety*

$$\mathbf{Hilb}^{r,d}(E/S) \xrightarrow{\pi} S$$

projective over S. The fibre of $\mathbf{Hilb}^{r,d}(E/S)$ over a point s is isomorphic to the Hilbert scheme $\mathbf{Hilb}^{r,d}(E(s))$.

Proof Since S is assumed to be affine, we can find, over $S \times X$, an exact sequence of vector bundles

$$0 \longrightarrow N \longrightarrow S \times \mathcal{O}_X(-n)^k \xrightarrow{j} E \longrightarrow 0.$$

Let $E_0 = \mathcal{O}_X(-n)^k$ so that the map j induces an embedding

$$\mathbf{Hilb}^{r,d}(E/S, S') \longrightarrow \mathrm{Mor}(S', S) \times \mathbf{Hilb}^{r,d}(E_0, S')$$

which has image given by pairs $(\alpha, \mathbf{F}')$ such that the composite map $\sigma_{\alpha,\mathbf{F}'} : \alpha^*(N) \to \mathbf{F}'$ on $S' \times X$ vanishes. Let $\sigma : N \to \mathbf{F}$ denote the map associated to the universal quotient sheaf $\mathbf{F}$ on $\mathbf{Hilb}^{r,d}(E_0) \times X$. We can assume that n is chosen large enough so that $H^1(X, Q(n)) = 0$ for all quotients Q of rank r and degree d of one of the bundles $E(s)$. This is possible because the family is bounded. Then $\mathrm{Ext}^1(N(s), \mathbf{F}(s)) = 0$. Let

$$\mathrm{pr}_{12} : S \times \mathbf{Hilb}^{r,d}(E_0) \times X \longrightarrow S \times \mathbf{Hilb}^{r,d}(E_0)$$

be the projection onto the first two factors. Then the coherent sheaf on $S \times \mathbf{Hilb}^{r,d}(E_0)$ given by $\mathrm{pr}_{12*}(\mathcal{H}om(N, \mathbf{F}))$ is locally-free. The map σ can be identified with a section of this vector bundle and if $\beta : S' \to \mathbf{Hilb}^{r,d}(E_0)$ is the map associated to the family $\mathbf{F}'$ parametrized by S' then we have $\sigma_{\alpha,\mathbf{F}'} = (\alpha, \beta)^*(\sigma)$. This section vanishes if and only if the map $(\alpha, \beta) : S' \to S \times \mathbf{Hilb}^{r,d}(E_0)$ factors through the scheme of zeros of σ. This vanishing scheme gives us the variety which represents the functor. □

Proposition 8.6.4 *Let* $q = (s, F) \in \mathbf{Hilb}^{r,d}(E/S)$ *with* $F = E(s)/G$. *Then we have an exact sequence*

$$0 \longrightarrow \mathrm{Hom}(G, F) \longrightarrow T_q\,\mathbf{Hilb}^{r,d}(E/S) \xrightarrow{T_q\pi} T_sS \xrightarrow{\omega} \mathrm{Ext}^1(G, F).$$

Proof The kernel of the differential of π at the point $q = (s, F)$ is obviously the tangent space $T_q\,\mathbf{Hilb}^{r,d}(E(s))$. This gives us the first part of the exact sequence. To define the second map, we consider a tangent vector $t \in \mathrm{Mor}(D, S) = T_sS$. Let I be the maximal ideal of the double point. As in the proof of Theorem 8.2.1, the problem of extending the quotient $F = E(s)/G$ to $D \times X$ has a solution if and only if the class $\omega(t) \in \mathrm{Ext}^1(G, F \otimes I)$ given by the image of the canonical extension in $\mathrm{Ext}^1(E(s), E(s) \otimes I)$ via the diagram

$$\begin{array}{ccccccccc} & & & & & & G & & \\ & & & & & & {\scriptstyle i}\downarrow & & \\ 0 & \longrightarrow & E(s) \otimes I & \longrightarrow & E_D & \longrightarrow & E(s) & \longrightarrow & 0 \\ & & {\scriptstyle p\otimes\mathrm{id}}\downarrow & & & & & & \\ & & F \otimes I & & & & & & \end{array}$$

vanishes. Since $I = \mathbb{C}\epsilon$, this gives a class $\omega(t) \in \mathrm{Ext}^1(G, F)$ and so we get a map $t \mapsto \omega(t) : T_sS \to \mathrm{Ext}^1(G, F)$ which is linear in t because, as in Lemma 8.2.3, it can be identified with the value at t of the image in $\mathrm{Ext}^1(G, F \otimes T_s^*)$ of the canonical element $\kappa \in \mathrm{Ext}^1(E(s), E(s) \otimes T_s^*)$ given by the extension

$$0 \longrightarrow E(s) \otimes T_s^* \longrightarrow E_{S_1} \longrightarrow E(s) \longrightarrow 0,$$

where (S_1, s) is the infinitesimal neighbourhood of order 1 of s in S and T_s is the tangent space to S at s. This exact sequence can be considered as an exact sequence of $\mathcal{O}_X$-modules via the projection $S_1 \times X \to X$ to the second factor. □

Definition 8.6.5 *The map* $\kappa : T_sS \to \mathrm{Ext}^1(E(s), E(s))$ *defined above is called the* Kodaira-Spencer infinitesimal deformation map *of* E *at the point* s.

The map $\omega : T_sS \to \mathrm{Ext}^1(G, F)$ is therefore given by $\omega = p \circ \kappa \circ i$, where $i : G \to E(s)$ and $p : E(s) \to F$ are the canonical maps.

Proof of the existence theorems

Let F be a vector bundle of rank r and degree d on X. We can assume that d is sufficiently large so that F is generated by its sections and that $H^1(F) = 0$, so that if we consider a vector space H of dimension $\chi = d + r(1-g)$, the choice of an isomorphism $\xi : H \simeq H^0(F)$ allows us to view $q = (\xi, F)$ as a point of the Hilbert scheme $\mathbf{Hilb}^{r,d}(H \otimes \mathcal{O}_X)$. It follows from Theorem 8.2.1 that this Hilbert scheme is smooth in a neighbourhood of this point. Take a smooth affine open neighbourhood S of s and consider the universal quotient bundle $\mathbf{F}$ on $S \times X$. Put $E = H \otimes \mathcal{O}_X$.

Lemma 8.6.6 *Let s be a point of S and $G(s) = \ker(E \to F(s))$. Then the Kodaira-Spencer infinitesimal deformation map κ can be identified with the boundary map $T_s = \mathrm{Hom}(G(s), \mathbf{F}(s)) \to \mathrm{Ext}^1(\mathbf{F}(s), \mathbf{F}(s))$ associated to the exact sequence*

$$0 \longrightarrow G(s) \longrightarrow E \longrightarrow \mathbf{F}(s) \longrightarrow 0$$

by applying the functor $\mathrm{Hom}(-, \mathbf{F}(s))$.

It follows that the map κ is surjective at the point s. For $r'' > 0$, we consider the relative Hilbert scheme

$$\mathbf{Hilb}^{r'',d''}(\mathbf{F}/S) \xrightarrow{\pi^{r'',d''}} S$$

of the quotients $F'' = \mathbf{F}(s)/F'$. The map $\omega : T_s \to \mathrm{Ext}^1(F', F'')$ of Proposition 8.6.4 is still surjective and that proposition shows that, at the point q'' given by the quotient F'', we have

$$\operatorname{rk} T_{q''}\pi^{r'',d''} = \dim_s S - \dim \mathrm{Ext}^1(F', F'').$$

Put $\mu'' = d''/r''$ and $\mu = d/r$. To see that $\mathbf{M}(r,d)$ (respectively, $\mathbf{M}^s(r,d)$) is not empty it suffices to prove that the union of the images of the maps $\pi^{r'',d''}$, for $\mu'' < \mu$ (respectively $\leq$) is not S: for the points s which are not contained in this union, the bundle $F(s)$ is semi-stable (respectively stable). We note that this union is finite and closed because such a quotient must be a quotient of the trivial bundle and so $\mu'' \geq 0$. We know by Bertini that $\dim \operatorname{Im} \pi^{r'',d''} \leq \operatorname{rk} T_{q''}\pi^{r'',d''}$. So it suffices to observe that under the hypothesis on the slope, we have $\dim \mathrm{Ext}^1(F', F'') > 0$ for all quotients F'' in one of the relative Hilbert schemes above. But, by the Riemann-Roch formula,

$$\dim \mathrm{Ext}^1(F', F'') \geq -\chi(\mathcal{H}om(F', F'')) = r'r''(\mu' - \mu'' + g - 1),$$

where r' and μ' denote the rank and slope of F', respectively. If $\mu'' < \mu$ then the right hand side is strictly positive for all $g \geq 1$ which shows that $\mathbf{M}(r,d)$ is not empty. If $g \geq 2$ then we obtain strict inequality for $\mu'' \leq \mu$ which shows that $\mathbf{M}^s(r,d)$ is non-empty. □

Proof of Lemma 8.6.6 Let G be the universal sub-bundle and (S_1, s) the infinitesimal neighbourhood of order 1 of the point s in S. Consider the diagram

$$\begin{array}{ccccccccc} & & 0 & & 0 & & 0 & & \\ & & \downarrow & & \downarrow & & \downarrow & & \\ 0 & \longrightarrow & G(s)\otimes T_s^* & \longrightarrow & G_{S_1} & \longrightarrow & G(s) & \longrightarrow & 0 \\ & & \downarrow & & \downarrow & & \downarrow & & \\ 0 & \longrightarrow & E\otimes T_s^* & \longrightarrow & E_{S_1} & \longrightarrow & E & \longrightarrow & 0 \\ & & \downarrow & & \downarrow & & \downarrow & & \\ 0 & \longrightarrow & \mathbf{F}(s)\otimes T_s^* & \longrightarrow & \mathbf{F}_{S_1} & \longrightarrow & \mathbf{F}(s) & \longrightarrow & 0 \\ & & \downarrow & & \downarrow & & \downarrow & & \\ & & 0 & & 0 & & 0 & & \end{array}$$

of short exact sequences of sheaves on X. The sequence for E splits canonically and this splitting induces a map $G(s) \to \mathbf{F}_{S_1}$ which allows us to construct a map of short exact sequences

$$\begin{array}{ccccccccc} 0 & \longrightarrow & G(s) & \longrightarrow & E & \longrightarrow & \mathbf{F}(s) & \longrightarrow & 0 \\ & & {\scriptstyle\rho}\downarrow & & \downarrow & & \| & & \\ 0 & \longrightarrow & \mathbf{F}(s)\otimes T_s^* & \longrightarrow & \mathbf{F}_{S_1} & \longrightarrow & \mathbf{F}(s) & \longrightarrow & 0 \end{array}$$

The map ρ is the one which gives, by transposition, the identification $T_s \simeq \mathrm{Hom}(G(s), \mathbf{F}(s))$ (cf. section 8.2). Then we can write down the derived exact sequences of the functor $\mathrm{Hom}(-, T_s^* \otimes \mathbf{F}(s))$, giving the following commuting diagram:

$$\begin{array}{ccc} \mathrm{Hom}(G(s), \mathbf{F}(s)\otimes T_s^*) & \xrightarrow{\delta} & \mathrm{Ext}^1(\mathbf{F}(s), \mathbf{F}(s)\otimes T_s^*) \\ \uparrow & & \| \\ \mathrm{Hom}(\mathbf{F}(s)\otimes T_s^*, \mathbf{F}(s)\otimes T_s^*) & \longrightarrow & \mathrm{Ext}^1(\mathbf{F}(s), \mathbf{F}(s)\otimes T_s^*) \end{array}$$

in which δ is the boundary map. Taking the image of the identity of $\mathbf{F}(s) \otimes T_s^*$ we see that $\delta(\rho)$ coincides up to sign with κ; but only up to sign because the image of the identity under the second boundary map is not the class associated to the extension but its opposite. This completes the proof of Lemma 8.6.6. □

8.7 Exercises

Consider an elliptic curve X. The object of the following problems is to describe the structure of the algebraic vector bundles of rank r and degree d on X. Recall that the canonical bundle ω_X is trivial.

The line bundles

(1.1) Let L be an invertible bundle of degree 1 on X. Show that $h^0(L) = 1$ and that a non-zero section vanishes at exactly one smooth point.
(1.2) Construct an isomorphism $X \to \mathrm{Pic}^1(X)$. Deduce that all of the components $\mathrm{Pic}^d(X)$ of the Picard group are isomorphic to X.

The unstable bundles

(2.1) Let F and F' be two semi-stable bundles of slopes μ and μ' respectively. Show that if $\mu < \mu'$ then we have $\mathrm{Ext}^1(F, F') = 0$.
(2.2) Deduce that any unstable bundle of rank r on X is a direct sum of semi-stable bundles. [Show, by induction on the rank, that the Harder-Narasimhan filtration splits.]

The case where r and d are coprime

Let $\mathbf{M}(r,d)$ denote the moduli space of S-equivalence classes of semi-stable bundles of rank r and degree d on X. In this section, r and d are assumed to be coprime so the moduli space consists of isomorphism classes of stable bundles of degree r and d. We assume that $d > 0$.
(3.1) Show that a bundle F on X is stable if and only if it has no endomorphisms other than homotheties.
(3.2) Let F be a stable bundle of rank r and degree d and let F^* be the dual of F. Show that $h^0(F) = d$ and that the choice of a generator of $H^0(\omega_X)$ gives us an isomorphism $H^1(F^*) \simeq H^0(F)^*$.
(3.3) Choose such a generator once and for all; this corresponds by duality to a generator of $H^1(\mathcal{O}_X)$. Show that we can construct an extension

$$0 \longrightarrow H^0(F) \otimes \mathcal{O}_X \xrightarrow{i} G \longrightarrow F \longrightarrow 0$$

such that the boundary map of the associated long exact sequence in cohomology can be identified with the identity map $H^0(F) \to H^0(F)$. Show that the map i factors through the evaluation map as follows:

$$\begin{array}{ccc} H^0(F) \otimes \mathcal{O}_X & & \\ {\scriptstyle j}\downarrow & \searrow^{i} & \\ H^0(G) \otimes \mathcal{O}_X & \xrightarrow[\mathrm{ev}]{} & G \end{array}$$

Also, show that the map j is an isomorphism.

(3.4) Show that the bundle G is stable of rank $r+d$ and degree d. [Note that if it were the direct sum of semi-stable bundles then the cokernel of the evaluation map would also be a direct sum and then use Exercise 2.1.]

(3.5) Deduce that there exists a map $\mathbf{M}(r,d) \to \mathbf{M}(r+d,d)$ making the diagram

$$\begin{array}{ccc} \mathbf{M}(r,d) & \longrightarrow & \mathbf{M}(r+d,d) \\ & {\scriptstyle \det}\searrow \quad \swarrow{\scriptstyle \det} & \\ & \mathrm{Pic}^d(X) & \end{array}$$

commute, where det denotes the determinant map.

(3.6) Deduce that the map $\det : \mathbf{M}(r,d) \to \mathrm{Pic}^d(X)$ is an isomorphism. [Induct on the rank.]

The arbitrary r and d case

Let ν denote the gcd of r and d and suppose that $\nu > 1$. The variety $\mathbf{M}(r,d)$ is therefore the space of equivalence classes of semi-stable bundles of rank r and degree d. Set $r' = r/\nu$ and $d' = d/\nu$.

(4.1) Show that there exists a finite morphism $\varphi : \mathbf{M}(r',d')^\nu \to \mathbf{M}(r,d)$. Deduce that $\dim \mathbf{M}(r,d) \geq \nu$.

(4.2) Deduce that there are no stable bundles of rank r and degree d.

(4.3) Show that the map φ is equivariant for the natural action of the symmetric group $\mathfrak{S}(\nu)$ on $\mathbf{M}(r',d')^\nu$ and that the induced map on the quotient

$$\mathbf{M}(r',d')^\nu/\mathfrak{S}(\nu) \longrightarrow \mathbf{M}(r,d)$$

bijects.

(4.4) Let $\mathrm{Div}^\nu(X)$ denote the Hilbert scheme of effective divisors of degree ν on the curve X, in other words, the Hilbert scheme $\mathbf{Hilb}^{0,\nu}(\mathcal{O}_X)$. Show that $\mathrm{Div}^\nu(X)$ is smooth and that the canonical map

$$X^\nu \longrightarrow \mathrm{Div}^\nu(X)$$

can be identified with the geometric quotient of X^ν under the action of the symmetric group.

(4.5) Show that the variety $\mathbf{M}(r,d)$ is smooth. [Use the fact that, in a neighbourhood of a point q represented by a polystable bundle $F = \bigoplus_{i=1}^{\nu} F_i$ with F_i stable of rank r' and degree d', the variety $\mathbf{M}(r,d)$ is, from an analytical viewpoint, locally isomorphic to the quotient

$$\mathrm{Ext}^1(F,F)/\,\mathrm{Aut}(F),$$

where the action of the group of automorphisms is by conjugation. This quotient can be identified with a known space.]

(4.6) Deduce that there is an isomorphism $\mathrm{Div}^\nu(X) \simeq \mathbf{M}(r,d)$.

[In the last two questions, §4 and §5 of chapter II of [57] may be helpful.]

Part II

Semi-stable Sheaves on the Projective Plane

Introduction

The object of the following chapters[1] is to give an insight into the techniques leading to the fundamental theorems, which are now well known, in the theory of moduli spaces of semi-stable sheaves of fixed rank and Chern class on the projective plane. These give conditions of existence [15] and irreducibility [17, 15], and allow us to compute the Picard group [14]. These results will be given in chapters 16, 17 and 18. They do not contain any new results, although the method used to compute the Picard groups is new and is inspired by an example which was studied in the Bowdoin paper to illustrate Drézet's Theorem [37]. This proof was only possible with the help of Göttsche and Hirschowitz who have proved a version of the Brill-Noether Theorem for the projective plane which is sufficient for the application which we have in mind. This proof was announced at the annual EUROPROJ conference in Catania.

In the proofs of all of these results the Shatz stratification of complete families of torsion-free sheaves associated to the Harder-Narasimhan filtration plays an important role. In fact, what we need is a formula to compute the codimension of the strata. This is done by studying the stratification from a differential point of view. We do this in chapter 15. The results which we present in chapters 11, 12 and 13 hold, by and large, for any surface and the proofs can, on the whole, be reduced to those we give for the projective plane. This is the case for the Restriction Theorem for semi-stable sheaves, for Bogomolov's Theorem or to show that the family of semi-stable sheaves on an algebraic surface is bounded. We also give a sketch of the construction of moduli spaces on

[1] This part formed a course in the workshop "Vector Bundles on Surfaces" organized by EUROPROJ (European Projective Geometry Network) and CIMI (Centre International pour les Mathématiques et l'Informatique) at Nice Sophia-Antipolis, 14–18 June, 1993.

all algebraic surfaces. In this construction we shall follow the method of Simpson [58] which applies to all algebraic varieties, including singular ones. It relies on a bound on the dimension of the vector space of sections of a semi-stable sheaf. We have obtained the estimates here in dimension 2 by using the theorem about restricting to a curve of sufficiently large degree.

Chapter 9 contains some general results about semi-stable sheaves on surfaces. As on curves, we define the Harder-Narasimhan filtrations for torsion-free coherent algebraic sheaves and the Jordan-Hölder filtrations for semi-stable sheaves. We also give some standard examples of semi-stable sheaves on the projective plane. Chapter 10 contains some additional material about bundles on curves: we show there that the tensor product of two semi-stable sheaves on a curve is still semi-stable. This is proved by calling on a result of Narasimhan and Seshadri [49] which states that each stable vector bundle on a curve comes from a unitary representation of the fundamental group. This result is vital in obtaining the Restriction to Curves Theorem. On higher dimensional varieties there exist analogous versions for μ-semi-stable bundles [44] which are useful to derive Flenner's Theorem for higher dimensional varieties but we will not give them here.

Finally, in this part we shall be using spectral sequences, notably in chapters 13 and 15, for which we refer the reader to Lang's book [35] or to Grothendieck's paper [28].

9

Filtrations

The notion of semi-stable sheaf on a smooth projective algebraic variety is inspired by the one we have given in chapter 5 for curves: we ask that the sheaf be torsion-free and we replace the slope by the reduced Hilbert polynomial. This notion was originally introduced by Gieseker for smooth projective surfaces [22]: he shows that the introduction of semi-stable sheaves enables us to compactify the moduli space of μ-stable bundles. This notion was taken up again by Maruyama [43] for higher dimensional varieties. We can still define Harder-Narasimhan filtrations for torsion-free sheaves and Jordan-Hölder filtrations for semi-stable sheaves which, as we have already seen for curves, are of tremendous importance in the description of moduli spaces.

9.1 Semi-stability

Semi-stable sheaves on a curve (see chapter 5)

Let F be a coherent algebraic sheaf of rank $r > 0$ and Chern class c_1 on a smooth irreducible projective curve C. The group $H^2(C, \mathbb{Z})$ is identified with $\mathbb{Z}$. The degree of F is the number $\deg(F)$ given by $\langle [C], c_1 \rangle$, where $[C]$ denotes the fundamental class of C. The slope of F is the ratio

$$\mu(F) = \frac{\deg(F)}{r}.$$

Definition 9.1.1 *Such a coherent algebraic sheaf F on C is said to be semi-stable (respectively stable) if*

(1) it is locally-free and

(2) for each coherent sub-sheaf $F' \subset F$ of rank $0 < r' < r$ we have $\mu(F') \leq \mu(F)$ (respectively $<$).

The semi-stable sheaves of fixed slope form an abelian category. The stable sheaves are simple objects in this category: this means that if F is a stable sheaf then $\mathrm{End}(F) = \mathbb{C}$.

Semi-stable sheaves on a surface

The notion of semi-stability and stability given above for curves extends to all projective varieties. However, we shall limit ourselves here to the case of surfaces.

Let X be a smooth connected projective algebraic surface which is equipped with a very ample sheaf $\mathcal{O}_X(1)$ of Chern class h in $H^2(X, \mathbb{Z})$. The group $H^4(X, \mathbb{Z})$ will be identified with $\mathbb{Z}$ and the group $H^2(X, \mathbb{Z})$ comes with an intersection form $(u, v) \mapsto u.v$. Given a coherent algebraic sheaf F on X of rank r and Chern classes c_1 and c_2, we let P_F denote the Hilbert polynomial of F. It is given by

$$P_F(m) = \frac{1}{2}m(m+1)r \deg X + \chi_1 m + \chi,$$

where $\chi = \sum_i (-1)^i h^i(F)$ is the Euler-Poincaré characteristic of F and $\chi_1 = \chi(F|_Y) = c_1.h + r(1-g)$ is the Euler-Poincaré characteristic of the restriction of F to a general hyperplane section $Y \in |\mathcal{O}_X(1)|$. The Euler-Poincaré characteristic is given by the Riemann-Roch formula

$$\chi = r\chi(\mathcal{O}_X) + \frac{1}{2}c_1(c_1 - \omega_X) - c_2,$$

where ω_X is the Chern class of the canonical bundle $\Lambda^2 T^*(X)$. Finally, $\deg(X) = h^2$.

The *degree* of F is the number $\deg(F) = c_1.h$. If $r > 0$ then we introduce the ratios

$$\mu(F) = \frac{\deg(F)}{r} \qquad \text{and} \qquad p_F = \frac{P_F}{r}$$

called the *slope* and *reduced Hilbert polynomial* of F, respectively.

Definition 9.1.2 *A coherent algebraic sheaf F of rank $r > 0$ on a surface X is said to be semi-stable (respectively stable) if*

(1) it is torsion-free and

(2) for each coherent sub-sheaf $F' \subset F$ of rank $0 < r' < r$ we have

$$p_{F'} \le p_F \text{ (respectively } <\text{)}.$$

In this definition, the polynomials are ordered lexicographically. So the last inequality means that $\mu(F') \leq \mu(F)$ and, in the case of equality,

$$\frac{\chi(F')}{r'} \leq \frac{\chi(F)}{r}.$$

If we replace the reduced Hilbert polynomial in this definition with the slope then we obtain the notion of μ-semi-stable and μ-stable sheaf. We have the following implications:

$$\begin{array}{ccc} \text{semi-stable} & \Longrightarrow & \mu\text{-semi-stable} \\ \Uparrow & & \\ \text{stable} & \Longleftarrow & \mu\text{-stable} \end{array}$$

Example.
On the projective plane, the Chern classes can be identified with numbers.

Proposition 9.1.3 *Let F be a rank 2 torsion-free sheaf on the projective plane with Chern classes $c_1 = 0$ and c_2.*

(1) If F is semi-stable and non-trivial then $c_2 \geq 2$ and F has no non-trivial sections.

(2) If F is locally-free and has no non-trivial sections then it is μ-stable.

Proof The semi-stability condition implies that for all coherent sub-sheaves $F' \subset F$, $c_1(F') \leq \mu(F) = 0$ and in the case of equality, $c_2(F') \geq \frac{1}{2}c_2(F)$. This condition implies in particular that $h^0(F(-1)) = 0$ and we also have $h^2(F(-1)) = 0$ because $H^2(F(-1))$ is dual to $\mathrm{Hom}(F, \mathcal{O}(-2))$ by Serre duality. Then, $\chi(F(-1)) = -c_2 \leq 0$. If $c_2 > 0$ then we cannot have $F' = \mathcal{O}$ as it contradicts $c_2(F') \geq \frac{1}{2}c_2(F)$. So $h^0(F) = 0$ and we also have $h^2(F) = 0$ by Serre duality. Since $\chi(F) = 2 - c_2$, we have $c_2 \geq 2$. If $c_2 = 0$ then, since $\chi(F) = 2$, there exists a trivial rank 1 sub-sheaf F' and the quotient F/F' is still a trivial sheaf of rank 1. It follows that F is itself trivial. This establishes (1). To show assertion (2) it suffices to remark that if F is locally-free then a rank 1 sub-bundle F' is forced to be contained in an invertible sub-sheaf of F. The condition $h^0(F) = 0$ implies that $h^2(F) = 0$ by Serre duality, and so $c_2 \geq 2$. This implies μ-stability because the second Chern class c_2 of an invertible sheaf F' is zero. □

Remark.
The same argument shows that there are no semi-stable sheaves of any rank with Chern classes $c_1 = 0$ and $c_2 = 1$ on the projective plane. On the other hand, it is easy to find μ-semi-stable locally-free sheaves of rank 2 with Chern classes $c_1 = 0$ and $c_2 = 1$. It suffices to consider the ideal $\mathcal{J}_a$ of a point $a \in \mathbf{P}_2$ and generic extensions

$$0 \longrightarrow \mathcal{O} \longrightarrow F \longrightarrow \mathcal{J}_a \longrightarrow 0.$$

These extensions are classified by $\mathrm{Ext}^1(\mathcal{J}_a, \mathcal{O}) = \mathbb{C}$. A non-trivial extension corresponds to a locally-free sheaf and such a sheaf is μ-semi-stable.

The following can be proved in the same way:

Proposition 9.1.4 *Let F be a rank 2 torsion-free coherent sheaf on the projective plane with Chern classes $c_1 = -1$ and c_2. Then $c_2 \geq 1$ and the following are equivalent:*

(1) the sheaf F is μ-semi-stable;

(2) the sheaf F is μ-stable.

Furthermore, if F is locally-free then these conditions are equivalent to the following condition:
(3) the sheaf F has no non-trivial sections.

9.2 The Harder-Narasimhan filtration

Let X be a surface. Given a coherent algebraic sheaf F on X, a point $x \in X$ such that F_x is not a free module is said to be *singular* for F. If F is torsion-free then there are only finitely many singular points of F. The degree of the coherent sub-sheaves $F' \subset F$ is then bounded: to see this, it suffices to restrict to a curve which does not pass through any of the singular points of F. Moreover, if we fix the degree of F' then the Euler-Poincaré characteristic of F' is also bounded. Hence, amongst the sub-sheaves of F we can find one whose reduced Hilbert polynomial is maximal. If we choose a sub-sheaf $F' \subset F$ from those of maximal rank then we obtain the notion of *maximal sub-sheaf* of F. This sub-sheaf is semi-stable and the quotient F/F' is still a torsion-free coherent sheaf for which we can repeat the above construction. We obtain

Proposition 9.2.1 *Every torsion-free coherent sheaf F has an increasing filtration by non-zero sub-sheaves*

$$\{0\} \subset F_1 \subset F_2 \subset \cdots \subset F_\ell = F$$

such that if p_i denotes the reduced Hilbert polynomial of $\mathrm{gr}_i = F_i/F_{i-1}$ *then we have*

(1) the sequence p_i is strictly decreasing,
(2) the quotient gr_i is semi-stable.

Such a filtration is unique.

This filtration is called the *Harder-Narasimhan filtration* of F. It generalizes the one which we have seen for curves in section 5.4 which was introduced by Harder and Narasimhan. The sequence of polynomials p_i are called the *Harder-Narasimhan polynomials* of F. We set $p_{\max} = p_1$ and $p_{\min} = p_\ell$. The slopes $\mu_i = \mu(\mathrm{gr}_i)$ form the *Harder-Narasimhan sequence* of F and we put $\mu_{\min} = \mu_1$ and $\mu_{\max} = \mu_\ell$.

Proposition 9.2.2 *Let F and G be two torsion-free coherent sheaves on the surface X. If* $\mathrm{Hom}(F, G) \neq 0$ *then we have* $p_{\min}(F) \leq p_{\max}(G)$.

As an example of the use of Harder-Narasimhan filtrations consider the following.

Lemma 9.2.3 *Let Z be the kernel of the canonical map*

$$H^0(\mathcal{O}_{\mathbf{P}_2}(d)) \otimes \mathcal{O}_{\mathbf{P}_2} \longrightarrow \mathcal{O}_{\mathbf{P}_2}(d).$$

Then Z is a semi-stable bundle of rank $\frac{1}{2}d(d+3)$.

Proof For $d = 1$, the kernel is a rank 2 vector bundle with Chern classes $(-1, 1)$ which we denote by Q^*. It has no non-trivial sections and so it is μ-stable. In the general case, we use the fact that the bundle Z is a homogeneous bundle under the action of GL(3). There exists a filtration of Z by an increasing sequence of homogeneous sub-bundles Z_i $(i = 0, \ldots, d-1)$ and the associated grading is $\mathrm{gr}_i = S^{d-i}Q^* \otimes \mathcal{O}(i)$. Each of these gradings is irreducible. The sub-bundle Z_i is the kernel of the epimorphism $H^0(\mathcal{O}_{\mathbf{P}_2}(d)) \otimes \mathcal{O}_{\mathbf{P}_2} \to H^0(\mathcal{O}_{\mathbf{P}_2}(d-i-1)) \otimes \mathcal{O}_{\mathbf{P}_2}(i+1)$. Moreover, the exact sequence

$$0 \longrightarrow Z_{i-1} \longrightarrow Z_i \longrightarrow \mathrm{gr}_i \longrightarrow 0$$

does not split: indeed, we also have the exact sequence

$$0 \to \mathrm{gr}_i \to H^0(\mathcal{O}(d-i)) \otimes \mathcal{O}(i) \to H^0(\mathcal{O}(d-i-1)) \otimes \mathcal{O}(i+1) \to 0$$

and a non-zero map $\mathrm{gr}_i \to Z_i \hookrightarrow H^0(\mathcal{O}_{\mathbf{P}_2}(d)) \otimes \mathcal{O}_{\mathbf{P}_2}$ would have to extend to $H^0(\mathcal{O}(d-i)) \otimes \mathcal{O}_{\mathbf{P}_2}(i)$ which is possible only if $i = 0$.

A non-zero homogeneous sub-bundle $G \subset Z$ must take the form Z_i. In fact, G contains Z_0 otherwise there exists an integer i such that $G \cap Z_{i-1} = \{0\}$ and $G \cap Z_i \neq \{0\}$ and $G \cap Z_i$ would then split the above exact sequence. If $Z_{i-1} \subset G$ and $Z_{i-1} \neq G$ then the quotient G/Z_{i-1} can be identified with a non-zero homogeneous sub-bundle of

$$H^0(\mathcal{O}_{\mathbf{P}_2}(d-i)) \otimes \mathcal{O}_{\mathbf{P}_2}(i)$$

and, after what we have said (replacing d by $d-i$), would contain $\mathrm{gr}_i = S^{d-i}Q^* \otimes \mathcal{O}_{\mathbf{P}_2}(i)$. But then this implies that $Z_i \subset G$. Hence, if i is the largest integer such that $Z_i \subset G$, we see that $Z_i = G$.

Now, if Z is not semi-stable then each of the terms of the Harder-Narasimhan filtration would be invariant under the action of GL(3) and so take the form Z_i. To complete the proof it suffices to show that $\mu(Z_i) < \mu(Z)$. We have $\mu(\mathrm{gr}_i) = (-d+3i)/2$. These slopes are strictly increasing and, hence, so are the slopes $\mu(Z_i)$. Thus, none of the sheaves Z_i destabilize Z and so Z is semi-stable. □

9.3 Jordan-Hölder filtrations

The semi-stable sheaves with given reduced Hilbert polynomial form an abelian category. In this abelian category, the increasing sequences and the decreasing sequences are stationary. It follows that if we are given a semi-stable sheaf F with reduced Hilbert polynomial p then we can find an increasing filtration

$$\{0\} \subset F_1 \subset F_2 \subset \cdots \subset F_\ell = F$$

such that $\mathrm{gr}_i = F_i/F_{i-1}$ are stable with reduced Hilbert polynomial p. Such a filtration is not, in general, unique but the sheaf $\mathrm{gr}(F) = \bigoplus_{i=1}^{\ell} \mathrm{gr}_i(F)$ is well defined up to isomorphism. Such a filtration is called a *Jordan-Hölder filtration* and the sheaf $\mathrm{gr}(F)$ is the *Jordan-Hölder grading*. Two semi-stable sheaves F and F' are said to be S-equivalent if their Jordan-Hölder gradings are isomorphic.

10

Operations on Semi-stable Bundles

The point of this chapter is to prove that the tensor product of two semi-stable bundles F and G on a smooth projective curve is still semi-stable and similarly for the symmetric powers S^pF and exterior powers Λ^pF. This result will be used in the following chapter in the proof of the theorem of the restriction of semi-stable sheaves to a curve. It is achieved by reducing to the case where the degree of the bundles is zero, by passing to a ramified cover and using the result of Narasimhan and Seshadri [49] which shows that, on a smooth curve, every stable bundle of degree zero comes from a unitary representation of the fundamental group. In higher dimensions Maruyama has obtained results of the same type for semi-stable sheaves [44]; this follows from the result for curves once we have established the Theorem of Restriction to Curves (cf. chapter 11).

10.1 Finite morphisms

Here we consider smooth connected projective curves C and C', and finite maps $f : C' \to C$.

Definition 10.1.1 *A finite map $f : C' \to C$ is called a* branched Galois cover *if there exists a finite group G acting on C' such that*

(i) the map f is G-equivariant,
(ii) the induced map $C'/G \to C$ is an isomorphism.

The group G can then be identified with the group of automorphisms φ of C' such that $f\varphi = f$: this is the *Galois group* of C'/C which we denote by $\mathrm{Gal}(C'/C)$.

Let C be a smooth connected algebraic projective curve and $K = \mathrm{Rat}(C)$ the field of rational functions of C. We know [11] that there exists an anti-equivalence of categories between the category of finite maps $f : C' \to C$, where C' is smooth and connected, and the category of finite extensions $K \hookrightarrow K'$. This correspondence is obtained by associating the extension $K \hookrightarrow K'$ to C' given by the field of rational functions of C'. Under this correspondence, the ramified Galois covers correspond to finite Galois extensions. Moreover, the Galois group $\mathrm{Gal}(C'/C)$ can be identified with the Galois group $\mathrm{Gal}(K'/K)$ of the extension.

Lemma 10.1.2 *Let C and C' be two smooth connected projective algebraic curves and let $f : C' \to C$ be a finite map. Then there exist a smooth connected projective curve Γ and a ramified Galois cover $g : \Gamma \to C$ which is factorized as in the following diagram:*

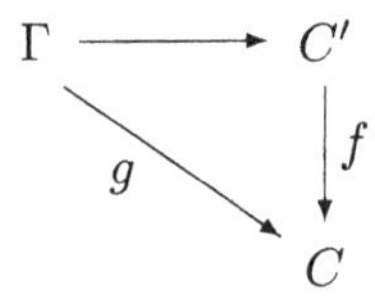

Proof It suffices to check that every finite extension $K \hookrightarrow K'$ embeds in a Galois extension. This is elementary (cf. [35]). □

Theorem 10.1.3 *Let C and C' be two smooth connected projective curves and $f : C' \to C$ a finite morphism. Let F be an algebraic vector bundle over C. Then F is semi-stable if and only if its inverse image $f^*(F)$ is semi-stable.*

Proof Let d be the degree of f. Suppose first that $f^*(F)$ is semi-stable. We have $\mu(f^*(F)) = d\mu(F)$ and for each vector sub-bundle $F' \subset F$ we also have $\mu(f^*(F')) = d\mu(F')$. Since $\mu(f^*(F')) \leq \mu(f^*(F))$, we get $\mu(F') \leq \mu(F)$.

For the converse, we can assume, by changing f if necessary using the above lemma, that the map $C' \to C$ is a ramified Galois cover. The Galois group $\mathrm{Gal}(C'/C)$ then acts on the set of sub-bundles of $f^*(F)$ and the maximal sub-bundle (i.e. the first term in the Harder-Narasimhan filtration) of $f^*(F)$ must be invariant under the Galois group. This means that the maximal sub-bundle descends to C. If F is semi-stable then this sub-bundle is F and so $f^*(F)$ is semi-stable. □

10.2 Tensors

We come to the main aim of this chapter:

Theorem 10.2.1 *Let E and F be two semi-stable vector bundles of slopes $\mu(F)$ and $\mu(G)$ on the curve C.*

(1) The tensor product $F \otimes G$ is semi-stable of slope

$$\mu(F \otimes G) = \mu(F) + \mu(G).$$

(2) The symmetric power S^pF and the exterior power Λ^pF are semi-stable of slope $p\mu(F)$.

Proof (1) We can reduce to the case where $\mu(F)$ and $\mu(G)$ are integers by pulling back if necessary to a ramified cover of C. Let $r(F)$ and $r(G)$ be the ranks of F and G. Then we can find rank one bundles L_F and L_G such that $L_F^{\otimes r(F)} = \det(F)$ and $L_G^{\otimes r(G)} = \det(G)$ and write

$$F = F' \otimes L_F \text{ and } G = G' \otimes L_G,$$

where F' and G' are semi-stable sheaves of degree 0. Since the notion of semi-stability on curves is invariant under tensoring with invertible bundles, we can reduce to the case where F and G have degree 0.

(2) In the case where F and G have degree 0 we can assume that they are stable by considering Jordan-Hölder filtrations.

Consider the fundamental group π of C and the universal cover $\tilde{C} \to C$. To each unitary representation $\rho : \pi \to U(r)$ of the fundamental group we can associate an analytic vector bundle

$$E_\rho = \tilde{C} \times_\pi \mathbb{C}^r$$

of rank r and degree 0, by taking the quotient of the trivial bundle $\tilde{C} \times \mathbb{C}^r$ by the natural action of the fundamental group: the fundamental group acts by $(g, (x, v)) \mapsto (xg^{-1}, gv)$, where $g \in \pi$ and $(x, v) \in \tilde{C} \times \mathbb{C}^r$. In view of the Serre Comparison Theorems, the bundle E_ρ is actually algebraic.

Theorem 10.2.2 (Narasimhan and Seshadri [49]) *(i) Let ρ be a unitary representation of the fundamental group of π. Then the associated vector bundle E_ρ over C is semi-stable.*

(ii) Let F be a stable bundle of degree 0 on a curve C. Then F gives rise to an irreducible representation of the fundamental group.

We shall assume this result, commenting only that part (i) can be easily obtained by calculating the slope of the sub-bundles of E_ρ in the following way: the natural Hermitian metric on E_ρ induces an Hermitian metric on each algebraic sub-bundle $F \subset E_\rho$. There exists a unique connection on F (cf. chapter 3) compatible with this Hermitian metric and the holomorphic structure of F. The curvature form of this connection allows us to compute the degree: the fact that the degree is negative can then be obtained by studying this curvature form. The same proof shows that if we are given an irreducible representation then the bundle E_ρ is stable.

Part (ii) is more delicate. The proof given by Narasimhan and Seshadri uses the moduli space of stable sheaves $\mathbf{M}^s(r,0)$ of semi-stable sheaves of rank r and degree 0 on the curve C (cf. chapter 7) and the moduli space $\mathrm{Rep}^0(\pi, \mathrm{U}(r))$ of conjugacy classes of unitary representations of the fundamental group; taking the connectivity of $\mathbf{M}^s(r,0)$ into account, the proof is reduced to two steps:

— a local study of the natural map

$$\mathrm{Rep}^0(\pi, \mathrm{U}(r)) \longrightarrow \mathbf{M}^s(r,0)$$

mapping the class of an irreducible representation ρ to the class of the stable vector bundle E_ρ. These two spaces are smooth (real) algebraic varieties and one can show that the linear tangent map bijects. It then follows that this map is open (in the usual topology).

— verify that the map is proper.

Another proof was given by Donaldson [10] which rests on the existence of global solutions of partial differential equations of heat equation type.

In our situation we see that the bundles F and G give rise to a representation of the fundamental group. But then this also follows for the tensor product $F \otimes G$, the symmetric powers $S^p F$ and the exterior powers $\Lambda^p F$ and so these are semi-stable. □

Corollary 10.2.3 *Let F and G be two vector bundles on the curve C. Then*

$$\mu_{\max}(F \otimes G) = \mu_{\max}(F) + \mu_{\max}(G).$$

Exercise

Let F be a semi-stable bundle of rank r on a curve C and $\rho : \mathrm{GL}(r, \mathbb{C}) \to \mathrm{GL}(m, \mathbb{C})$ a linear representation. Then the rank m vector bundle F_ρ

associated to F and to the representation ρ (cf. chapter 1) is still semistable.

11

Restriction to Curves

It is clear that if the restriction of a torsion-free coherent sheaf F to a general curve of the linear system $|\mathcal{O}_X(d)|$ is semi-stable then the sheaf F is itself μ-semi-stable. In this chapter, we shall prove the converse. The statement that we give here is a particular case of a result of Flenner [20] for varieties of arbitrary dimension. It generalizes the classical result first obtained by Grauert and Mülich [26] for rank 2 bundles and successively improved by Barth [5], Forster, Hirschowitz and Schneider [21], Spindler [60], Maruyama [44], Hirschowitz [21], Mehta and Ramanathan [46]. The proof given here is inspired both by the work of Hirschowitz [21] and by Flenner's article.

Theorem 11.0.1 *Let $(X, \mathcal{O}_X(1))$ be a polarized algebraic surface and F a μ-semi-stable sheaf of rank r. Let d be an integer such that*

$$d \geq \frac{r^2}{2} \deg X$$

and let Y be a general curve in the linear system $|\mathcal{O}_X(d)|$. Then $F|_Y$ is a semi-stable bundle.

11.1 The Shatz stratification

Let E be a vector bundle of rank r and degree d on a curve. Consider an increasing filtration

$$0 \subset F_1 \subset \cdots \subset F_\ell = E$$

by sub-sheaves of strictly increasing rank. Let (r_i, d_i) denote the rank and degree of the sub-sheaf F_i. The integer ℓ will be called the *length* of the filtration. To such a filtration we associate the piecewise linear function $\mathfrak{p}_F : [0,r] \to \mathbb{R}$ whose graph is the polygonal curve on $\mathbb{R}^2$ obtained by joining the points with coordinates (r_i, d_i) in the plane:

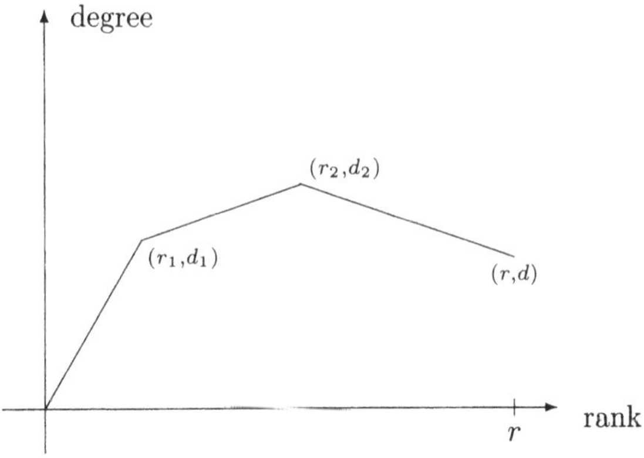

This polygonal curve has $\ell + 1$ vertices given by each of the terms F_i of the filtration. The natural order on the functions $[0,r] \to \mathbb{R}$ induces a preorder on the filtrations of E.

Lemma 11.1.1 *Let F be a maximal filtration of E. Then F is the Harder-Narasimhan filtration.*

Proof If F_i is a maximal filtration then the associated curve is forced to be concave which means that the grading gr_i satisfies the following condition:

$$i \mapsto \mu(\mathrm{gr}_i) \text{ is strictly decreasing.}$$

It remains to show that each grading gr_i is locally-free and semi-stable. We show that E/F_i is torsion-free. If this were not the case then we could find another filtration F'_i of the same length such that $F_i \subset F'_i$ for each i and where the sub-sheaves F_i and F'_i have the same rank. We then have $\mathfrak{p}_{F'} > \mathfrak{p}_F$ which contradicts the maximality. It follows that each of the terms gr_i is torsion-free and hence locally-free and semi-stable. So this filtration is the Harder-Narasimhan filtration. □

Now consider a proper smooth map $Y \to S$ of relative dimension 1 over a smooth connected algebraic variety S. Let E be a rank r vector bundle over Y. For each $s \in S$, we let Y_s denote the fibre over s and E_s the restriction of E to Y_s and the degree of E_s is denoted d (which is constant).

Consider the Hilbert scheme

$$\mathbf{Hilb}^{r_1,d_1}(E/S) \longrightarrow S$$

of pairs (s, F_1) given by a point $s \in S$ and a sub-sheaf $F_1 \subset E_s$ of rank r_1 and d_1. This scheme is relatively projective over S. More generally, given a polygonal curve $\mathfrak{p}$ with ℓ_1 vertices, we can consider the Hilbert scheme

$$\mathbf{Flag}(E/S, \mathfrak{p}) \longrightarrow S$$

of pairs given by a point $s \in S$ and a flag $F = (F_i)$,

$$0 \subset F_1 \subset \cdots \subset F_\ell = E_s,$$

such that $\mathfrak{p}_F = \mathfrak{p}$. This scheme is still proper over S. Consequently, the image of this scheme is a closed subset $S_\mathfrak{p}$ of S.

Proposition 11.1.2 *(1) There exists an non-empty open subset S' of S over which the Harder-Narasimhan polygon of E_s is constant.*

(2) Over this open set S', there exists an increasing filtration of E by sub-bundles $F_i \subset E$ such that, for all $s \in S'$, the induced filtration $F_{i,s}$ is the Harder-Narasimhan filtration of E_s.

Proof Consider the polygons $\mathfrak{p}$ such that $S_\mathfrak{p} = S$. Choose $\mathfrak{p}$ maximal amongst such polygons. Let S' be the complement of the union of closed sets $\bigcup S_{\mathfrak{p}'}$ with $\mathfrak{p}' > \mathfrak{p}$. This union is obviously a finite union and consequently, it too is closed. So S' is a non-empty open subset of S. Over the open set S', the bundle E_s has a filtration with polygon $\mathfrak{p}$ and this filtration is maximal: this filtration must be the Harder-Narasimhan filtration of E_s by the above lemma. Over S', the map

$$\mathbf{Flag}(E/S', \mathfrak{p}) \longrightarrow S'$$

is bijective and proper, hence finite. The variety S is assumed to be smooth and the normality property of S implies that there exists a section. By the universal property of the flag scheme, we obtain a filtration of E over S' by sub-sheaves which induce the Harder-Narasimhan filtration over each fibre. By flatness, the successive quotients are locally-free and the sub-sheaves are forced to be sub-bundles. □

Remark.
The same proof shows that the points $s \in S$ for which $\mathfrak{p}_{E_s} = \mathfrak{p}$ form a locally closed subset of S. This is the *Shatz stratification.*

11.2 The Harder-Narasimhan sequence of $F|_C$

In what follows, we consider a finite morphism $\pi : X \to \mathbf{P}_2$ such that $\pi^*(\mathcal{O}_{\mathbf{P}_2}(1)) = \mathcal{O}_X(1)$. This morphism is flat of degree $\deg(X)$. The family of curves of degree d on the plane gives, in the inverse image, a linear system of curves $R \subset |\mathcal{O}_X(d)|$ of dimension $\frac{1}{2}d(d+3)$. Theorem 11.0.1 follows from the following result.

Theorem 11.2.1 *Let d be an integer and C a general curve in the linear system $R \subset |\mathcal{O}_X(d)|$. If F is a μ-semi-stable sheaf on X then the Harder-Narasimhan filtration of $F|_C$ satisfies the following condition:*

$$\mu_i - \mu_{i+1} \leq \frac{2d}{d+1} \deg X.$$

The sheaf F has only a finite number of singularities; let E denote the restriction of F to the open set over which it is locally-free. Consider the family $S \subset R$ of smooth curves of R which do not pass through any of the singular points of F and let $Y \to S$ denote the universal curve over S. Then we have a diagram

$$\begin{array}{ccc} Y & \xrightarrow{p_2} & X \\ {\scriptstyle p_1}\downarrow & & \\ S & & \end{array}$$

Consider the bundle $\mathcal{E} = p_2^*(F)$ to which we apply Proposition 11.1.2. We can assume, by shrinking S if necessary, that the conclusions of that proposition are satisfied over S. Consider the sub-bundle F_i of $\mathcal{E}$. By the universal property of the relative Grassmannian scheme, we get a diagram

$$\begin{array}{ccc} Y & \xrightarrow{\quad\varphi\quad} & \mathbf{Grass}(r_i, E) \\ & \searrow \quad \swarrow & \\ & X & \end{array}$$

The map p_2 identifies Y with an open subset of a bundle of projective spaces over X and it has local sections. We will show that if

$$\mu_i - \mu_{i+1} > \frac{2d}{d+1} \deg X$$

then the sub-bundle F_i comes from a sub-bundle of E. To do this, it suffices to prove that φ remains constant along the fibres of the projection

p_2. For this, we show that the vertical differential

$$d\varphi : T(Y/X) \longrightarrow \mathcal{H}om(F_i, \mathcal{E}/F_i)$$

vanishes at all points of Y. Consider a curve $C \subset X$ in the linear system corresponding to a point $s \in S$. Over C, the point s defines a section of p_2. We can consider the restriction of $d\varphi$ along this section. We obtain a map of vector bundles over the curve C

$$T(Y/X)|_C \longrightarrow \mathcal{H}om(F_{i,s}, E_s/F_{i,s}),$$

where E_s is the restriction of E to the curve C. But we have

$$\mu_{\max}(\mathcal{H}om(F_{i,s}, E_s/F_{i,s})) = \mu_{i+1} - \mu_i.$$

To see that this differential vanishes, it suffices to establish that

$$\mu_{\min}(T(Y/X)|_C) > \mu_{i+1} - \mu_i.$$

Theorem 11.2.2 (Flenner) *If C is a general curve in the linear system R then we have*

$$\mu_{\min}(T(Y/X)|_C) \geq -\frac{2d}{d+1} \deg X.$$

We now complete the proof of Theorem 11.2.1. Suppose that $\mu_i - \mu_{i+1} > 2d \deg X/(d+1)$. Then by the above theorem, we see that

$$\mu_{i+1} - \mu_i < -\frac{2d}{d+1} \deg X \leq \mu_{\min}(T(Y/X)|_C).$$

Thus, the sub-bundle F_i comes from a sub-bundle of E defined in a neighbourhood of C. Obviously, this sub-bundle extends to a coherent sub-sheaf of F, still denoted by F_i, defined over the surface X itself, and we have

$$\mu(E|_C) = d\mu(E) \text{ and } \mu(F_{i,s}) = d\mu(F_i).$$

It follows that this sub-sheaf contradicts the μ-semi-stability of E. □

End of the proof of Theorem 11.0.1 We start with a lemma which tells us more about the Harder-Narasimhan sequence of the restriction to a general curve and which we shall find very useful in what follows.

Lemma 11.2.3 *Let C be a general curve in the linear system R and gr_i the Harder-Narasimhan grading of $E|_C$. Then the invertible bundle $\det \mathrm{gr}_i$ is the restriction of an invertible bundle on X to C.*

Proof Consider the linear system $R \subset |\mathcal{O}_X(d)|$ and, as before, we let $Y \to R$ denote the universal curve over R. This is a bundle of projective spaces over X and so its Picard group is $\mathrm{Pic}(Y) = \mathrm{Pic}(X) \oplus \mathbb{Z}\ell$, where ℓ is the generator of $\mathrm{Pic}(R) = \mathbb{Z}$. The isomorphism is induced by the embedding $Y \hookrightarrow R \times X$. The pairs $(s, x) \in Y$, where s represents a curve missing a given point x_0, form an open set U which is the complement of a hyperplane section coming from R. Hence, the restriction map $\mathrm{Pic}(X) \to \mathrm{Pic}(U)$ surjects. If we take x_0 to be a point which is not contained in the curve C then the invertible bundle $\det \mathrm{gr}_i$ extends to U and so comes from X. The lemma follows from this. □

Suppose that $F|_C$ is not semi-stable. Then there are at least two terms in the Harder-Narasimhan filtration and we have $0 < \mu_i - \mu_{i+1} \leq 2d \deg X/(d+1)$. Let r_i be the rank of gr_i. By the above lemma, we can write $\mu_i = \alpha_i d/r_i$, where $\alpha_i \in \mathbb{Z}$, and the inequality above leads to

$$0 < \frac{\alpha_i}{r_i} - \frac{\alpha_{i+1}}{r_{i+1}} \leq \frac{2}{d+1} \deg X,$$

$$\frac{4}{r^2} \leq \frac{2}{d+1} \deg X.$$

This implies $d + 1 \leq r^2 \deg X/2$ which contradicts the hypothesis $d \geq r^2 \deg X/2$. □

Proof of Theorem 11.2.2 The curve C comes from a degree d curve Γ in $\mathbf{P}_2$. The tangent space to R at the point s defined by Γ can be identified with $H^0(N_\Gamma)$, where $N_\Gamma = \mathcal{O}_\Gamma(d)$ is the normal bundle to Γ in the projective plane. The vertical tangent space $T(Y/X)|_C$ is the inverse image under the projection $\pi : C \to \Gamma$ of the kernel K of the evaluation map

$$H^0(N_\Gamma) \otimes \mathcal{O}_\Gamma \xrightarrow{\mathrm{ev}} N_\Gamma.$$

Recall Lemma 9.2.3 which states that the kernel Z of the canonical map

$$H^0(\mathcal{O}_{\mathbf{P}_2}(d)) \otimes \mathcal{O}_{\mathbf{P}_2} \longrightarrow \mathcal{O}_{\mathbf{P}_2}(d)$$

is a semi-stable bundle.

Lemma 11.2.4 *Let K be the kernel of the evaluation map*

$$H^0(N_\Gamma) \otimes \mathcal{O}_\Gamma \xrightarrow{\mathrm{ev}} N_\Gamma.$$

Then, for general Γ, we have

$$\mu_{\min}(K) \geq -\frac{2d}{d+1}.$$

Proof We have an exact sequence

$$0 \longrightarrow \mathcal{O}_\Gamma \longrightarrow Z|_\Gamma \longrightarrow K \longrightarrow 0$$

and $\mathcal{O}_\Gamma$ is the first term of the Harder-Narasimhan filtration of $Z|_\Gamma$: so we have $\mu_{\min}(K) = \mu_{\min}(Z|_\Gamma)$. We have seen that the Harder-Narasimhan filtration of the restriction of a vector bundle to a general curve extends to the universal curve $Y \to S$ of X. Then the determinant of each gr_i is in the $\mathrm{Pic}(X)$ factor. If we apply this to the case $X = \mathbf{P}_2$ and to the bundle Z then we obtain, in particular, that the determinant of the last Harder-Narasimhan quotient comes from $\mathbf{P}_2$: $\det \mathrm{gr}_\ell = \mathcal{O}_\Gamma(t)$ for a suitable integer t. If p is the rank of gr_ℓ then we have

$$\mu_{\min}(K) = \frac{td}{p}$$

and we have

$$-1 \geq t \geq -\frac{2(p+1)}{d+3}.$$

Indeed, the former inequality is obvious since the degree of $Z|_\Gamma$ is negative. For the latter, observe that $\mathrm{Hom}(\Lambda^p Z, \mathcal{O}_\Gamma(t)) \neq 0$ and we compute $\mathrm{Hom}(\Lambda^p Z, \mathcal{O}_\Gamma(t))$ using the exact sequences

$$0 \longrightarrow \mathcal{O}(-d) \longrightarrow \mathcal{O} \longrightarrow \mathcal{O}_\Gamma \longrightarrow 0, \tag{11.1}$$

$$0 \longrightarrow \Lambda^{p+1} Z \longrightarrow \Lambda^{p+1} H \otimes \mathcal{O} \longrightarrow \Lambda^p Z(d) \longrightarrow 0, \tag{11.2}$$

where we have put $H = H^0(\mathcal{O}(d))$. We obtain the following diagram of exact sequences:

$$\begin{array}{ccccc}
 & & & & \mathrm{Hom}(\Lambda^{p+1} Z, \mathcal{O}_{\mathbf{P}_2}(t)) \\
 & & & & \downarrow \\
\mathrm{Hom}(\Lambda^p Z, \mathcal{O}_{\mathbf{P}_2}(t)) & \to & \mathrm{Hom}(\Lambda^p Z, \mathcal{O}_\Gamma(t)) & \to & \mathrm{Ext}^1(\Lambda^p Z, \mathcal{O}_{\mathbf{P}_2}(t-d)) \\
 & & & & \downarrow \\
 & & & & 0
\end{array}$$

But we have the following lemma:

Lemma 11.2.5 *The exterior powers of the bundle Z are semi-stable.*

As one of the terms $\mathrm{Hom}(\Lambda^p Z, \mathcal{O}_{\mathbf{P}_2}(t))$ and $\mathrm{Hom}(\Lambda^{p+1} Z, \mathcal{O}_{\mathbf{P}_2}(t))$ must be non-zero we get

$$\frac{(p+1)d}{N-1} + t \geq 0,$$

where $N = h^0(\mathcal{O}_{\mathbf{P}_2}(d)) = \frac{1}{2}(d+1)(d+2)$. This leads to the stated lower bound and these inequalities lead to $2p \geq d+1$. So we have

$$\mu_{\min}(K) \geq -d\Big(1 + \frac{1}{p}\Big)\frac{2}{d+3} \geq -\frac{2d}{d+1},$$

which proves Lemma 11.2.4. □

Proof of Lemma 11.2.5 For each integer p, we obviously have the lower bound

$$\mu_{\min}(\Lambda^p Z) \geq -d$$

by the exact sequence (11.2) above, but the desired result is more precise. It suffices to find a curve $\Gamma' \subset \mathbf{P}_2$ over which the bundle Z restricts to a semi-stable bundle and then apply Theorem 10.2.1 to this curve. Consider a curve of degree $d' \geq d$ and the last term gr_ℓ of the Harder-Narasimhan filtration of $Z|_{\Gamma'}$. If the curve Γ' is sufficiently general then the invertible bundle $\det \mathrm{gr}_\ell$ comes from the plane

$$\det \mathrm{gr}_\ell = \mathcal{O}_{\Gamma'}(t)$$

and we have $t < 0$. Then we obtain a morphism $\omega : \Lambda^p Z \longrightarrow \mathcal{O}_{\Gamma'}(t)$, where $p = r_\ell$ is the rank of this grading, and this surjects along Γ'. The restriction map

$$\mathrm{Hom}(\Lambda^p Z, \mathcal{O}(t)) \longrightarrow \mathrm{Hom}(\Lambda^p Z, \mathcal{O}_{\Gamma'}(t)) \tag{11.3}$$

then has kernel $\mathrm{Hom}(\Lambda^p Z, \mathcal{O}(t-d'))$ and its cokernel embeds into

$$\mathrm{Ext}^1(\Lambda^p Z, \mathcal{O}(t-d')) \simeq \mathrm{Hom}(\Lambda^{p+1} Z, \mathcal{O}(t+d-d')).$$

So, if we choose d' large enough ($d' \geq 2d$ will suffice), then the above restriction map is an isomorphism. This means, in particular, that ω extents uniquely to the projective plane. Consider the induced map

$$\varphi : Z \longrightarrow \Lambda^{p-1} Z^*(t) \hookrightarrow \Lambda^p H^* \otimes \mathcal{O}(t+d).$$

This map has rank p along Γ' which implies that the exterior power $\Lambda^{p+1}\varphi$ vanishes on the curve Γ'. Set $t' = (p+1)(t+d)$. If d' is large enough ($d' \geq (N+1)d$ will suffice) then the restriction map analogous to (11.3)

$$\mathrm{Hom}(\Lambda^{p+1}Z, \mathcal{O}(t')) \longrightarrow \mathrm{Hom}(\Lambda^{p+1}Z, \mathcal{O}_{\Gamma'}(t')) \tag{11.4}$$

injects. So we get

$$\Lambda^{p+1}\varphi = 0$$

which means that the map φ has rank at most p. Then the image of φ is a rank p torsion-free coherent sheaf which extends gr_ℓ and has slope $\mu(\mathrm{Im}\,\varphi) \leq \mu(Z)$. Since Z is semi-stable, we obtain $Z = \mathrm{Im}\,\varphi$ and hence, $Z|_{\Gamma'} = \mathrm{gr}_\ell$. Thus, $Z|_{\Gamma'}$ is semi-stable. □

Remark.
The argument used in the above lemma is close to that used by Mehta and Ramanathan. It allows us to show, without a descent argument, that the restriction of any μ-semi-stable sheaf to a general curve in the linear system $|\mathcal{O}_X(d)|$ is still semi-stable provided that d is large enough. However, it does not allow us to compute a suitable bound on the degree d.

12

Bogomolov's Theorem

Let X be a smooth connected projective surface. Recall that the Néron-Severi group $\mathrm{NS}(X)$ of X is the kernel of the canonical map

$$H^2(X, \mathbb{Z}) \longrightarrow H^2(X, \mathcal{O}_X).$$

The Chern class c_1 of each algebraic vector bundle is contained in $\mathrm{NS}(X)$. Conversely, we know by a classical theorem due to Schwarzenberger [53], that if we are given an integer $r \geq 2$, a class $c_1 \in \mathrm{NS}(X)$ and an integer $c_2 \in \mathbb{Z}$ then there exists an algebraic vector bundle on X with rank r and Chern classes c_1 and c_2. If we ask that F be semi-stable then we need to impose further conditions on the Chern classes of F.

12.1 The discriminant of semi-stable sheaves

More generally, we can consider a coherent algebraic sheaf F of rank r and Chern classes c_1 and c_2 on a smooth projective polarized surface $(X, \mathcal{O}_X(1))$. We set

$$\nu = \frac{c_1}{r}$$

in $H^2(X, \mathbb{Q})$ and

$$\Delta = \frac{1}{2r^2}\left(2rc_2 - (r-1)c_1^2\right).$$

This rational number is called the *discriminant* of F. The discriminant satisfies the following properties:

(1) If F is an invertible bundle then we have $\Delta = 0$.
(2) If F' and F'' are vector bundles with discriminants Δ' and Δ'', respectively, then we have

$$\Delta(F' \otimes F'') = \Delta' + \Delta''.$$

In fact, this formula holds if only one of the sheaves F' or F'' is locally-free. In particular, Δ is invariant when we tensor F by an invertible bundle.

(3) Let F be a coherent algebraic sheaf of rank r and Chern classes c_1 and c_2. The Riemann-Roch formula can be written

$$\chi = r(P(\nu) - \Delta),$$

where $P(\nu) = \chi(\mathcal{O}_X) + \frac{1}{2}(\nu^2 - \nu.\omega_X)$.

Theorem 12.1.1 (Bogomolov [8]) *Let F be a semi-stable sheaf of rank $r > 0$ and discriminant Δ on a surface X. Then $\Delta \geq 0$.*

First, recall the following result (cf. Lemma 7.1.2).

Lemma 12.1.2 *Let F be a semi-stable bundle of rank r and slope μ on a smooth projective curve Y. Then*

$$\frac{h^0(F)}{r} \leq \begin{cases} \mu + 1 & \text{if } \mu \geq 0, \\ 0 & \text{if } \mu < 0. \end{cases}$$

Note that ω_X is the dualizing sheaf of X and put $k = \omega_X.h$ for the following.

Lemma 12.1.3 *Let F be a μ-semi-stable sheaf of slope 0 and rank $r > 0$. Suppose that d is an integer satisfying the following conditions:*

(1) there exists a curve $Y \in |\mathcal{O}_X(d)|$ such that $F|_Y$ is semi-stable, and

(2) $d.\deg X > k$.

Then the Euler-Poincaré characteristic of F satisfies the inequality

$$\frac{\chi(F)}{r} \leq \begin{cases} dk + 2 & \text{if } k \geq 0, \\ 1 & \text{if } k < 0. \end{cases}$$

Proof We have the exact sequence $0 \to F(-Y) \to F \to F|_Y \to 0$ and by semi-stability, $F(-Y)$, which has negative slope, has no non-trivial sections. It follows from the above lemma applied to the vector bundle $F|_Y$ that

$$\frac{h^0(F)}{r} \leq 1.$$

Now consider $h^2(F)$. By Serre duality, $h^2(F) = \dim \mathrm{Hom}(F, \omega_X)$. The sheaf $G = \mathcal{H}om(F, \omega_X)$ is still μ-semi-stable of slope k and its restriction to Y is also semi-stable. The sheaf $G(-Y)$ has slope $k - d \deg X < 0$. The argument we have already used shows that

$$\frac{h^2(F)}{r} \leq \begin{cases} dk+1 & \text{if } k \geq 0, \\ 1 & \text{if } k < 0. \end{cases}$$

Since $\chi(F) = \sum_i (-1)^i h^i(F)$, we obtain the desired inequality. □

Proof of Theorem 12.1.1 The bidual F^{**} is still μ-semi-stable and we have

$$\Delta(F) \geq \Delta(F^{**}).$$

Since the bidual is locally-free, we can then reduce the theorem to locally-free sheaves.

Suppose first that F is locally-free with Chern class $c_1 = 0$. We can choose a curve $Y \in |\mathcal{O}_X(d)|$ satisfying the conditions of the lemma above. Then the sheaf $F^{\otimes \ell}$ is still a μ-semi-stable locally-free sheaf of Chern class $c_1 = 0$ because this holds for the restriction to the curve Y. We can therefore apply the conclusion of the lemma to this vector bundle: taking account of the fact that $\Delta(F^{\otimes \ell}) = \ell \Delta(F)$ we obtain

$$\chi(\mathcal{O}_X) - \ell\Delta \leq \sup(dk+2, 1)$$

for each $\ell > 0$ which implies that $\Delta \geq 0$.

If c_1 is not zero then we can apply the above result to the endomorphism bundle $G = \mathcal{H}om(F, F)$ which remains μ-semi-stable but has Chern class $c_1 = 0$ and discriminant $2\Delta(F)$. This proves the result. □

Corollary 12.1.4 *Let* $\text{sign}(X)$ *be the signature of* X *and* $k = \omega_X.h$. *Suppose that* F *is a semi-stable sheaf on* X *of rank* r, *slope* μ *and Euler-Poincaré characteristic* χ. *Then*

$$\frac{\chi}{r} \leq \frac{1}{2\deg(X)}\left(\mu - \frac{k}{2}\right)^2 - \frac{1}{8}\text{sign}(X).$$

Proof By the Riemann-Roch formula and Bogomolov's Theorem, we have

$$\frac{\chi}{r} = P(\nu) - \Delta \leq P(\nu).$$

By definition

$$P(\nu) = \frac{1}{2}\left(\nu - \frac{\omega_X}{2}\right)^2 - \frac{1}{8}\text{sign}(X).$$

We know by the Index Theorem that this intersection form is negative on the orthogonal complement of h in $\mathrm{NS}(X)\otimes\mathbb{R}$. The product of $\nu-\omega_X/2$ with $h/\sqrt{h^2}$ is

$$\frac{\mu - k/2}{\sqrt{h^2}}$$

and so

$$\left(\nu - \frac{\omega_X}{2}\right)^2 \leq \frac{1}{\deg(X)}\left(\mu - \frac{k}{2}\right)^2$$

from which the corollary clearly follows. □

12.2 Simpson's estimate

For the construction of the moduli spaces of semi-stable sheaves, it is important to dispose of an analogous upper bound for $h^0(F)/r$. One such bound has been obtained by Simpson (in arbitrary dimensions). This bound generalizes the one we have found for curves in Lemma 12.1.2.

For all real numbers x, we set $[x]_+ = \sup(x, 0)$ as in chapter 7.

Theorem 12.2.1 (Simpson) *Let r be a positive integer. There exists a constant C depending only on r and the degree of X such that, for all semi-stable sheaves F of rank r on X, we have the following upper bound:*

$$\frac{h^0(F)}{r} \leq \frac{1}{2\deg(X)}([\mu + C]_+)^2.$$

Proof We use the fact that the function $x \mapsto [x]_+$ is an increasing function. Choose a general curve $Y \in |\mathcal{O}_X(1)|$ and consider the Harder-Narasimhan filtration of the restriction $F|_Y$ and its associated grading $\bigoplus_{i=1}^{\ell} \mathrm{gr}_i$. Let r_i and μ_i be the rank and slope of the grading gr_i. Then, by Lemma 12.1.2, we have

$$h^0(F|_Y) \leq \sum_i h^0(\mathrm{gr}_i) \leq \sum_i r_i[\mu_i + 1]_+$$

and by Theorem 11.2.1, we have $\mu_i - \mu_{i+1} \leq \deg(X)$. Hence, $\mu_i \leq \mu + r\deg(X)$. It follows that

$$\frac{h^0(F|_Y)}{r} \leq [\mu + r\deg(X) + 1]_+.$$

We apply this upper bound to the sheaves $F(-j)$ which have the same rank and slope $\mu - j\deg(X)$. Since F is torsion-free, $h^0(F(-j)) = 0$ for large enough j and the same applies to its restriction to the curve Y. Then we have

$$\frac{h^0(F)}{r} \leq \sum_{j=0}^{\infty} \frac{h^0(F(-j)|_Y)}{r}.$$

The above inequality implies

$$\frac{h^0(F)}{r} \leq \sum_{j=0}^{\infty} [\mu + (r-j)\deg(X) + 1]_+.$$

Comparing the above sum with an integral, we obtain

$$\frac{h^0(F)}{r} \leq \frac{1}{2\deg(X)}([\mu + C]_+)^2$$

with $C = (r+1)\deg(X) + 1$, which completes the proof. □

A careful examination of the above proof clearly shows that the constant C can be slightly improved.

13

Bounded Families

A family $\mathcal{F}$ of coherent sheaves on a smooth projective algebraic surface X is said to be bounded if there exist an algebraic variety S and a coherent algebraic sheaf $\mathcal{G}$ on $S \times X$ such that all sheaves F of the family are isomorphic to one of the sheaves $\mathcal{G}_s$ for $s \in S$. The aim of this chapter is to show that the family of semi-stable sheaves with fixed Hilbert polynomial on a smooth projective algebraic surface is bounded. The proof is in two steps: we show first that the result (in fact, a slightly stronger result) holds for the projective plane using Beilinson monads. In the case of an arbitrary smooth projective surface X we reduce to the first case by considering a finite map $X \to \mathbf{P}_2$.

13.1 The projective plane case

Theorem 13.1.1 *Let H be a polynomial and C a constant. The family $\mathcal{F}$ of torsion-free coherent sheaves with Hilbert polynomial H on the projective plane which satisfy the condition*

$$\mu_{\max}(F) \leq C$$

is bounded.

The proof rests on the following lemma:

Lemma 13.1.2 *Let H be a polynomial. The family of semi-stable sheaves with Hilbert polynomial H on the projective plane is bounded.*

We shall first show how the theorem follows from this lemma:

Proof of Theorem 13.1.1 Since the Hilbert polynomial of the sheaves F of the family is fixed, the condition $\mu_{\max}(F) \leq C$ implies that $\mu_{\min}(F)$ is bounded below and so the slope $\mu_i = \mu(\mathrm{gr}_i)$ of the sheaves which appear in the Harder-Narasimhan grading of the sheaves of $\mathcal{F}$ is also bounded. By the lemma, it suffices to show that the Hilbert polynomials of the sheaves gr_i are finite in number. Hence, we must prove that the discriminant $\Delta_i = \Delta(\mathrm{gr}_i)$ takes only a finite number of values. But by the Riemann-Roch formula, if $r_i = r(\mathrm{gr}_i)$ then we have

$$\sum_i r_i(P(\mu_i) - \Delta) = \chi(F)$$

and consequently the sum $\sum_i r_i \Delta_i$ is bounded. But we know from Bogomolov's Theorem that $\Delta_i \geq 0$. Hence, the Δ_i are bounded. □

Proof of Lemma 13.1.2 Fixing a Hilbert polynomial on the projective plane amounts to fixing the rank r and the Chern classes c_1 and c_2. We can then assume that the slope μ satisfies the following condition:

$$-1 < \mu \leq 0.$$

If F is not the trivial bundle of rank r then the Hodge numbers $h^0(F(i))$ are zero for $i \leq 0$. By Serre duality, the vector space $H^2(F(i))$ is dual to $\mathrm{Hom}(F, \mathcal{O}(-i-3))$ and so vanishes for $i \geq -2$. For $i = -2, -1, 0$ the only cohomology spaces which could be non-zero are the spaces $H^1(F(i))$ and their dimensions, computed via the Riemann-Roch formula, are given by

$$\begin{aligned} h^1(F) &= -\chi, \\ h^1(F(-1)) &= c_1 + r - \chi, \\ h^1(F(-2)) &= 2c_1 + r - \chi. \end{aligned}$$

Let Q^* denote the canonical sub-bundle of rank 2 on the projective plane defined by the kernel of the evaluation map

$$H^0(\mathcal{O}_{\mathbf{P}_2}(1)) \otimes \mathcal{O}_{\mathbf{P}_2} \longrightarrow \mathcal{O}_{\mathbf{P}_2}(1).$$

The fibre Q^*_x over a point $x \in \mathbf{P}_2$ can be identified with the space of sections of $\mathcal{O}_{\mathbf{P}_2}(1)$ which vanish at x.

Lemma 13.1.3 (The Beilinson monad) *Let F be a semi-stable sheaf of rank r and Chern classes (c_1, c_2) on the projective plane. Suppose that $-r < c_1 \leq 0$. Let $H_i = H^1(F(i-1))$ for $i = -1, 0, 1$ and consider the natural complex of locally-free sheaves*

$$0 \longrightarrow H_{-1} \otimes \Lambda^2 Q^* \xrightarrow{u} H_0 \otimes Q^* \xrightarrow{v} H_1 \otimes \mathcal{O}_{\mathbf{P}_2} \longrightarrow 0.$$

Then u is generically injective, v is generically surjective and the cohomology sheaf $\ker v/\operatorname{Im} u$ *is isomorphic to F.*

On the projective plane fixing the rank and the Chern classes or the Hilbert polynomial comes down to the same thing. Thus, this lemma obviously implies that the family of semi-stable sheaves with fixed Hilbert polynomial on the projective plane is bounded which completes the proof of the lemma. □

Proof The proof rests on the resolution of the diagonal of $\mathbf{P}_2 \times \mathbf{P}_2$. Consider the external tensor product $Q \boxtimes \mathcal{O}(1)$ on this product. If V is the space of sections of $\mathcal{O}(1)$ then the space of sections of Q can be identified with the dual V^* of V. Then the vector bundle $E = Q \boxtimes \mathcal{O}(1)$ has a canonical section whose zero scheme can be identified with the diagonal. So we can write down the Koszul resolution $K_\bullet$ (cf. [31]) of the structure sheaf of the diagonal Δ:

$$0 \longrightarrow \Lambda^2 E^* \longrightarrow E^* \longrightarrow \mathcal{O}_{\mathbf{P}_2 \times \mathbf{P}_2} \longrightarrow 0.$$

Consider the projections

$$\begin{array}{ccc} \mathbf{P}_2 \times \mathbf{P}_2 & \xrightarrow{p_2} & \mathbf{P}_2 \\ {\scriptstyle p_1}\downarrow & & \\ \mathbf{P}_2 & & \end{array}$$

The map $p_2 : \Delta \to \mathbf{P}_2$ is flat so the complex $K_\bullet \otimes p_2^*(F)$ is still a left resolution of $\mathcal{O}_\Delta \otimes p_2^*(F) = \mathcal{O}_\Delta \otimes \mathcal{O}_{\mathbb{P}_2} F$. Taking the direct image under p_1, the spectral sequence of derived functors gives rise to a spectral sequence (called the *Beilinson spectral sequence* [7]) whose $E_1^{p,q}$ term is given by

$$\begin{aligned} E_1^{p,q} &= R^q p_{1*}(\Lambda^{-p} Q^* \boxtimes F(p)) \\ &= \Lambda^{-p} Q^* \otimes H^q(F(p)) \end{aligned}$$

and which abuts on F in degree 0 and on 0 in the non-zero degrees. Taking account of the conditions we have already established for the Hodge numbers $h^q(F(i))$ with $i = -2, -1, 0$, this spectral sequence degenerates at the E_1 level to give the above complex, from which the result follows. □

13.2 The arbitrary surface case

Theorem 13.2.1 *Let X be a smooth projective algebraic surface. Let H be a polynomial and C a constant. The family $\mathcal{F}$ of torsion-free coherent* *algebraic sheaves on X with Hilbert polynomial H and such that*

$$\mu_{\max}(F) \leq C \tag{13.1}$$

is bounded.

This result is due to Maruyama [45].

Proof Consider a finite map

$$\pi : X \to \mathbf{P}_2$$

such that $\pi^*(\mathcal{O}_{\mathbf{P}_2}(1)) = \mathcal{O}_X(1)$. If F is a torsion-free coherent algebraic sheaf on X then its direct image $G = \pi_*(F)$ is still a torsion-free coherent sheaf on the projective plane and its Hilbert polynomial is still H. We are going to show that this sheaf still satisfies (13.1), possibly after changing the constant C. By Theorem 13.1.1 it is enough to show that the family of direct images is bounded.

Lemma 13.2.2 *There exists a constant C' such that, for all sheaves F of $\mathcal{F}$, we have*

$$\mu_{\max}(G) \leq C'$$

for the direct image $G = \pi_*(F)$.

Consider the algebra $\mathcal{A} = \pi_*(\mathcal{O}_X)$. The map π is flat because it is a finite morphism and the surface X is smooth. This implies that $\mathcal{A}$ is a locally-free sheaf on $\mathbf{P}_2$ of rank $\deg(X)$. Then we can introduce the rational numbers $\mu_{\min}(\mathcal{A})$ and $\mu_{\max}(\mathcal{A})$. On the other hand, there is a bijective correspondence between the set of coherent sub-sheaves of F and the set of coherent sub-$\mathcal{A}$-modules of G; this correspondence is given by the direct image and so preserves the Hilbert polynomial.

Proof of Lemma 13.2.2 Suppose that G_1 is the first term of the Harder-Narasimhan filtration of G. By definition this is a semi-stable sub-sheaf of G such that $\mu(G_1) = \mu_{\max}(G)$. Then the image G'_1 of the canonical map

$$G_1 \otimes_{\mathcal{O}_{\mathbf{P}_2}} \mathcal{A} \longrightarrow G$$

is a coherent sub-$\mathcal{A}$-module of G so this is the direct image of a sub-sheaf F_1 of F under π. The coherent sheaf F_1 has the same Hilbert polynomial as G'_1 and so the hypothesis of the lemma implies that $\mu(G'_1) \leq C$. Consider the torsion-free coherent sheaf given by $G_1 \otimes \mathcal{A}$ over the algebra

$\mathcal{O}_{\mathbf{P}_2}$. Since G_1' is a quotient of $G_1 \otimes \mathcal{A}$ we have the bounds $\mu_{\min}(G_1 \otimes \mathcal{A}) \leq \mu(G_1') \leq C$. Now, since the sheaf G_1 is semi-stable, we clearly have

$$\mu_{\min}(G_1 \otimes \mathcal{A}) = \mu(G_1) + \mu_{\min}(\mathcal{A})$$

which we can see by restricting to a curve with sufficiently large degree. So we get $\mu(G_1) \leq C - \mu_{\min}(\mathcal{A})$. □

Taking Theorem 13.1.1 into account, this implies that the family of direct images G we have constructed is bounded. It remains to show that the family of sheaves F is itself bounded. But it suffices to remark that such a sheaf F amounts to a sheaf G on $\mathbf{P}_2$ and an $\mathcal{A}$-module structure on G; that is, a map $\nu : \mathcal{A} \otimes G \to G$ of coherent algebraic sheaves on the projective plane which satisfies the following conditions:

(a) if $m : \mathcal{A} \otimes \mathcal{A} \to \mathcal{A}$ is the multiplication defined by the algebra structure on $\mathcal{A}$ then we have

$$\nu \circ (\mathrm{id}_{\mathcal{A}} \otimes \nu) = \nu(m \otimes \mathrm{id}_G)$$

in $\mathrm{Hom}(\mathcal{A} \otimes \mathcal{A} \otimes G, G)$,

(b) we have $\nu \circ (1 \otimes \mathrm{id}_G) = \mathrm{id}_G$ in $\mathrm{Hom}(G, G)$.

So we can parametrize the $\mathcal{A}$-module structures on G by the closed sub-scheme of $\mathrm{Hom}(\mathcal{A} \otimes G, G)$ given by the inverse image of the point $(\mathrm{id}_G, 0)$ in the map

$$\mathrm{Hom}(\mathcal{A} \otimes G, G) \longrightarrow \mathrm{Hom}(G, G) \times \mathrm{Hom}(\mathcal{A} \otimes \mathcal{A} \otimes G, G)$$

which associates to ν the pair of maps

$$\big(\nu \circ (1 \otimes \mathrm{id}_G), \nu \circ (\mathrm{id}_{\mathcal{A}} \otimes \nu) - \nu(m \otimes \mathrm{id}_G)\big)$$

of coherent algebraic sheaves.

Consider a reduced algebraic variety S and a coherent algebraic sheaf $\mathcal{G}$ on $S \times \mathbf{P}_2$ such that all sheaves G given as above are isomorphic to one of the sheaves $\mathcal{G}_s$. We can assume that the sheaf $\mathcal{G}$ is S-flat. Then by the Semi-continuity Theorem we can stratify S by a finite number of locally closed sub-varieties S_i over which the dimensions of the vector spaces $\mathrm{Hom}(\mathcal{G}_s, \mathcal{G}_s)$, $\mathrm{Hom}(\mathcal{A} \otimes \mathcal{G}_s, \mathcal{G}_s)$ and $\mathrm{Hom}(\mathcal{A} \otimes \mathcal{A} \otimes \mathcal{G}_s, \mathcal{G}_s)$ remain constant. Let $S' = \coprod_i S_i$. Then there exists a closed sub-scheme of a vector bundle $\mathcal{V} \to S'$ (of variable rank) which parametrizes the $\mathcal{A}$-module structure on the sheaves G. This shows that the family of sheaves F is bounded.

□

14

The Construction of the Moduli Space

We shall only give an outline of the construction. For more details the reader could consult Simpson's original article [58] or [38]. In fact, the construction is not much harder in all dimensions but it requires a generalization of the estimate given in Theorem 12.2.1. The proof is quite similar to the one given in chapter 7 for curves; however, the statement which generalizes the semi-stability criterion in section 7.1 must be read with care.

14.1 The coarse moduli space $M_X(H)$

Let X be a polarized projective algebraic surface and H a degree 2 polynomial. For any variety S, consider the set $\underline{\mathbf{M}}_X(H)(S)$ of isomorphism classes of S-flat families of semi-stable sheaves on X with Hilbert polynomial H parametrized by the algebraic variety S.

Theorem 14.1.1 *There exists a coarse moduli space* $\mathbf{M}_X(H)$ *for the functor* $\underline{\mathbf{M}}_X(H)$. *This space has the following properties:*

(i) The algebraic variety $\mathbf{M}_X(H)$ *is projective.*

(ii) The points of $\mathbf{M}_X(H)$ *are the S-equivalence classes of semi-stable sheaves with Hilbert polynomial H.*

(iii) The set of isomorphism classes of stable sheaves can be identified with an open subset of the variety $\mathbf{M}_X(H)$.

The coarse moduli property of $\mathbf{M}_X(H)$ means that there exists a functorial map $f : \underline{\mathbf{M}}_X(H)(S) \to \mathrm{Mor}(S, \mathbf{M}_X(H))$ called the *moduli morphism* satisfying the following universal property: for all other algebraic varieties $\mathbf{M}'$ and all functorial maps $g : \underline{\mathbf{M}}_X(H)(S) \to \mathrm{Mor}(S, \mathbf{M}')$, there exists one and only one map $\varphi : \underline{\mathbf{M}}_X(H) \to \mathbf{M}'$ making the following diagram commute:

$$\begin{array}{ccc} \underline{\mathbf{M}}_X(P)(S) & \xrightarrow{f} & \mathrm{Mor}(S, \mathbf{M}_X(H)) \\ & \searrow^{g} & \downarrow \varphi \\ & & \mathrm{Mor}(S, \mathbf{M}') \end{array}$$

Obviously this property characterizes the variety $\underline{\mathbf{M}}_X(H)$. This statement, first proved by Gieseker [22] for surfaces, was established for all polarized smooth projective varieties by Maruyama [43] and afterwards by Simpson [58] for sheaves of dimension d. Simpson's construction is the most natural.

The proof rests on Mumford's geometric invariant theory. First, we recall the essential definitions (see Chapter 6).

14.2 Invariant geometry (cf. chapter 6)

Let G be an affine algebraic group acting on an algebraic variety X. We recall the following definitions from section 6.1.

Definition 14.2.1 *A map $f : X \to Y$ of algebraic varieties is called a* good quotient *of X by G if f is G-equivariant and satisfies the following conditions:*

(i) the map f is affine and surjective,

(ii) the canonical map of algebras $\mathcal{O}_Y \to f_*(\mathcal{O}_X)^G$ is an isomorphism,

(iii) the image of a closed G-invariant subset is closed in Y and the map f separates the disjoint G-invariant closed subsets.

It is easy to see that a good quotient $f : X \to Y$ satisfies the universal property that we expect quotients to possess and so such a quotient, if it exists, is unique. Moreover, we can see that each fibre of f contains exactly one closed orbit; it follows that the set of closed points of Y can be identified with the set of closed orbits.

Definition 14.2.2 *A good quotient is called a* geometric quotient *if the orbits of G are closed in X.*

Examples

(a) Suppose that G is a reductive group. Let $X = \mathrm{Spec}(A)$ be an affine variety given by an algebra A of finite type. Then A^G is an algebra of finite type and the canonical morphism

$$\mathrm{Spec}(A) \longrightarrow \mathrm{Spec}(A^G)$$

associated to the inclusion $A^G \hookrightarrow A$ is a good quotient.

(b) (see section 6.4) Let G be a reductive group operating linearly on a vector space W and let $\mathbf{P}(W)$ denote the projective space of hyperplanes of W. Then this is a (geometric) quotient of $W^* - \{0\}$ by $\mathbb{C}^*$.

Definition 14.2.3 *A point $x \in \mathbf{P}(W)$ is said to be* semi-stable *under the action of G if there exists a G-invariant homogeneous polynomial on W^* which does not vanish at x. The points $x \in \mathbf{P}(W)$ which are semi-stable under the G action form an open set P^{ss}. A point $x \in \mathbf{P}(W)$ is said to be* stable *if it is semi-stable and if the map $g \mapsto gx : G \to P^{ss}$ is proper.*

We can now construct a projective variety Z and a good quotient $P^{ss} \to Z$. This variety is defined as follows: let $Y = W^*/G$ be the good affine quotient defined above and let 0 still denote the image of 0 in the quotient. Let W^{*ss} denote the inverse image of the complement of 0 in Y. This is an open $\mathbb{C}^*$-invariant subset. The group $\mathbb{C}^*$ acts properly on the open set $Y \setminus \{0\}$ and gives rise to a geometric quotient $Z = Y \setminus \{0\}/\mathbb{C}^*$. This variety is a projective variety and the composite map $W^{*ss} \to Z$ passes to the quotient to give a commuting diagram

$$\begin{array}{ccc} W^{*ss} & \longrightarrow & Y \setminus \{0\} \\ \downarrow & & \downarrow \\ P^{ss} & \xrightarrow{f} & Z \end{array}$$

The map f we have just found is the map we were looking for, making Z into a good quotient of P^{ss}. The stable points form an open subset P^s of P^{ss} on which the group G acts properly. This open subset is the inverse image of an open subset Z^s of Z and the map $P^s \to Z^s$ is a geometric quotient.

(c) In the situation of (b), consider a closed sub-scheme $X \subset \mathbf{P}(W)$ which is invariant under the action of G defined by a G-invariant ideal I of the graded algebra $A = \mathrm{Sym}(W)$ of polynomials on W^*. Then the homogeneous ideal I^G of the graded algebra A^G defines a closed sub-scheme $T \subset Z$ and the induced morphism

$$X^{ss} = X \cap P^{ss} \to T$$

is a good quotient of X^{ss} by G.

Mumford's criterion

There exists a useful criterion due to Mumford to determine which points are semi-stable and which are the stable points of $\mathbf{P}(W)$ under the action of G. Let $\lambda : \mathbb{C}^* \to G$ be a one parameter sub-group of G. We can then write

$$W^* = \bigoplus_{n \in \mathbb{Z}} W_n,$$

where W_n is the vector sub-space of linear forms $v \in W^*$ such that $\lambda(t).v = t^n v$ for all $t \in \mathbb{C}^*$. A point $x \in \mathbf{P}(W)$ can be represented by a linear form $v = \sum_n v_n$ with $v_n \in W_n$. Let $\mu(\lambda, x) = -\inf\{n \,|\, v_n \neq 0\}$.

Theorem 14.2.4 *There is a correspondence*

*x is semi-stable (respectively stable) $\Leftrightarrow$ for all λ, $\mu(\lambda, x) \geq 0$ (*resp. > 0*).*

This criterion allows us, for example, to establish the following results.

Example 1

Let E be a finite dimensional vector space. Consider the Grassmannian $\mathbf{Grass}(E \otimes \mathbb{C}^k)$ of r-dimensional vector space quotients of $E \otimes \mathbb{C}^k$ on which there is a natural action of the reductive group $\mathrm{SL}(E)$. This Grassmannian embeds via the Plücker embedding into the projective space $\mathbf{P}(\bigwedge^r(E \otimes \mathbb{C}^k))$, and this embedding is equivariant for the natural induced action of $\mathrm{SL}(E)$ on the vector space $\bigwedge^r(E \otimes \mathbb{C}^k)$. The following result is another version of Proposition 6.6.1.

Proposition 14.2.5 *The semi-stable points for the natural action of* $\mathrm{SL}(E)$ *on the Grassmannian* $\mathbf{Grass}(E \otimes \mathbb{C}^k)$ *correspond to the vector space quotients F of $E \otimes \mathbb{C}^k$ satisfying the following condition: for each non-trivial vector sub-space $E' \subset E$, the image F' of $E' \otimes \mathbb{C}^k$ in F satisfies the inequality*

$$\frac{\dim F'}{\dim E'} \geq \frac{\dim F}{\dim E}.$$

Example 2

Let $E = \mathcal{O}_X^r$ be the trivial bundle of rank r on an irreducible projective variety X and H a polynomial. Consider the Hilbert-Grothendieck scheme (cf. [29]) $\mathbf{Hilb}(E, H)$ of coherent quotient sheaves $E \to F$ with Hilbert polynomial $P_F = H$. If ℓ is a sufficiently large integer then the space of sections $H^0(F(\ell))$ has dimension $H(\ell)$ and is a quotient of $H^0(E(\ell))$ so that we obtain the morphism of varieties

$$\varphi_\ell : \mathbf{Hilb}(E, H) \hookrightarrow \mathbf{Grass}\big(H^0(E(\ell)), H(\ell)\big)$$

associating the quotient vector space $H^0(F(\ell))$ to a quotient sheaf F. In fact, this morphism is a closed embedding: this is the way Grothendieck constructed the Hilbert scheme. The map is equivariant for the natural action of $\mathrm{SL}(r) \subset \mathrm{Aut}(E)$ and the linear action of $\mathrm{SL}(r)$ on $H^0(E(\ell))$ enables us to consider the semi-stable points as in the preceding example. We then obtain

Proposition 14.2.6 *If ℓ is chosen large enough, the semi-stable points of $\mathbf{Hilb}(E, H)$ under the action of $\mathrm{SL}(r)$ are the points which represent quotients $E \to F$ satisfying the following condition: for each non-zero trivial sub-bundle $E' \subset E$ of rank r' with image F' in E', we have*

$$\frac{P_{F'}}{r'} \geq \frac{P_F}{r}.$$

Note that the open set of semi-stable points does not depend on ℓ providing that ℓ is chosen to be sufficiently large. A particular case of this proposition was given in Lemma 7.2.2.

14.3 The semi-stability criterion

We now need to express the notion of semi-stability in terms of sections and to have a result which allows us to describe the coherent sub-sheaves of a given semi-stable sheaf with the same reduced Hilbert polynomial. Let H be a polynomial. For each integer m consider the set $S_m(H)$ of isomorphism classes of torsion-free coherent sheaves on X with Hilbert polynomial H and which satisfy the following conditions.

(i) $H(m) \leq h^0(F(m))$.

(ii) For each non-zero coherent sub-sheaf $F' \subset F$ of rank r' we have

$$\frac{h^0(F'(m))}{r'} \leq \frac{h^0(F(m))}{r}.$$

On the other hand, consider the set $S_m^*(H)$ of isomorphism classes of torsion-free coherent sheaves F with Hilbert polynomial H which satisfy a sort of dual condition:

(ii)* For each torsion-free coherent quotient $F \to F''$ of rank r'', we have
$$\frac{H(m)}{r} \leq \frac{h^0(F''(m))}{r''}.$$

It is clear that we have $S_m(H) \subset S_m^*(H)$ for all m. For large enough m we can say more:

Proposition 14.3.1 *For a fixed polynomial H, there exists a number m_0 such that for all $m \geq m_0$, the sets $S_m(H)$ and $S_m^*(H)$ coincide with the set of isomorphism classes of semi-stable sheaves with Hilbert polynomial H.*

Remark.
One can ask if it would not be more natural simply to introduce the set of isomorphism classes of torsion-free coherent sheaves with Hilbert polynomial H which only satisfy condition (ii). On curves, condition (i) is effectively automatic. However, in dimension 2 this condition does not imply semi-stability even if we choose m very large.

We come now to the description of the Jordan-Hölder filtrations. Given a rank r semi-stable sheaf F and an integer m, we let $\mathrm{Max}_m(F)$ denote the set of non-zero coherent sub-sheaves $F' \subset F$ of rank r' which satisfy the following condition:
$$\frac{h^0(F'(m))}{r'} = \frac{h^0(F(m))}{r}.$$

Proposition 14.3.2 *Let H be a polynomial. There exists a number m_0 such that, for all $m \geq m_0$ and all semi-stable sheaves F with Hilbert polynomial H, the set $\mathrm{Max}_m(F)$ coincides with the set $\mathrm{Max}(F)$ of coherent sub-sheaves $F' \subset F$ having the same reduced Hilbert polynomial as F.*

The proofs of these two statements can be constructed in essentially the same way as that of Serre's Theorem B, once one establishes the fact that certain families of sheaves and sub-sheaves are bounded. This is the case, for example, for the family $\bigcup_{m \geq m_0} S_m^*(H)$ for sufficiently large m_0. To obtain this we use Theorem 13.1.1 and the estimate of Theorem 12.2.1 which allows us to give a lower bound on the slope $\mu_{\min}(F)$ of a sheaf of this family.

14.4 The construction

To prove Theorem 14.1.1 we choose m sufficiently large so that the following properties hold:

(a) for all semi-stable sheaves F with Hilbert polynomial H, the sheaf $F(m)$ is generated by its sections and $H^q(F(m)) = 0$ for $q > 0$, and the same also holds for all sub-sheaves $F' \in \mathrm{Max}(F)$;
(b) the conclusions of 14.3.1 and 14.3.2 hold.

Consider the vector bundle $E = \mathbb{C}^{H(m)} \otimes \mathcal{O}_X(-m)$ whose automorphism group can be identified with $\mathrm{GL}(H(m))$. Each semi-stable sheaf F with Hilbert polynomial H equipped with an isomorphism $\mathbb{C}^{H(m)} \simeq H^0(F(m))$ defines a point of the Hilbert scheme $\mathbf{Hilb}(E, H)$ and the induced map $H^0(E(m)) \to H^0(F(m))$ is an isomorphism. Proposition 14.2.6 applied to the Hilbert scheme and the criteria above allow us to see easily that such a point is semi-stable under the action of the group $\mathrm{SL}(H(m))$. The converse is more delicate because it is not obvious that a point of $\mathbf{Hilb}(E, P)$ which is semi-stable for the $\mathrm{SL}(H(m))$ action defines a semi-stable quotient sheaf nor that it is torsion-free, and we need to restrict to the (scheme theoretic) closure $\mathbf{Hilb}_0(E, H)$ of the open subset of torsion-free quotients. This difficulty can also be found in Gieseker's construction or in that of Maruyama.

Theorem 14.4.1 *Let $q \in \mathbf{Hilb}_0(E, H)$ define a coherent quotient sheaf $E \to F$. The following are equivalent:*

(i) the sheaf F is semi-stable and the canonical map

$$H^0(E(m)) \longrightarrow H^0(F(m))$$

is an isomorphism;

(ii) the point q is semi-stable under the action of $SL(H(m))$.

Let $\mathbf{Hilb}^{ss}(E, H)$ be the open subset of $\mathbf{Hilb}(E, H)$ of points which satisfy (i) above. The moduli space is the Mumford quotient

$$\mathbf{M}_X(H) = \mathbf{Hilb}^{ss}(E, H)/\mathrm{SL}(H(m)).$$

By construction, this is the Mumford quotient of semi-stable points of a closed sub-variety of $\mathbf{Hilb}(E,H)$: therefore, this is a projective variety and it is not too difficult, using the universal property of Hilbert schemes, to prove that it is a coarse moduli space and to identify the closed orbits. These are given by the quotients $E \to F$ with F *polystable*, in other words, a direct sum

$$F = \bigoplus_i F_i,$$

where F_i are stable with the same reduced Hilbert polynomial. This leads to Theorem 14.1.1.

14.5 Smooth points

Theorem 14.5.1 *Let F be a stable sheaf on X such that*

$$\mathrm{Ext}^2(F,F) = 0.$$

Then the moduli space $\mathbf{M}_X(H)$ is smooth in a neighbourhood of the point p defining F.

Proof By construction, using the notation we have used above, the open set of stable points $\mathbf{M}^s_X(H)$ of $\mathbf{M}_X(H)$ is the geometric quotient of the open set $\mathbf{Hilb}^s(E,H)$ of stable points under the action of the group $\mathrm{GL}(H(m))$. The stabilizer of a point of $\mathbf{Hilb}^s(E,H)$ can be identified with the automorphism group of the corresponding coherent sheaf F, that is, $\mathbb{C}^*$. It easily follows that the action of $\mathrm{GL}(H(m))$ factors through $\mathrm{PGL}(H(m))$ which acts properly and freely on the open set $\mathbf{Hilb}^s(E,H)$.

It therefore suffices to show that the Hilbert scheme $\mathbf{Hilb}^s(E,H)$ is smooth along the orbit corresponding to the point F. But we can turn to Grothendieck's criterion which allows us to decide if the Hilbert scheme is smooth in a neighbourhood of a point:

Theorem 14.5.2 *Let $F = E/F'$ be a point of the Hilbert scheme* $\mathbf{Hilb}(E,H)$.

(i) The Zariski tangent space to $\mathbf{Hilb}(E,H)$ *at the point F is isomorphic to* $\mathrm{Hom}(F',F)$.

(ii) If $\mathrm{Ext}^1(F',F) = 0$ *then the Hilbert scheme* $\mathbf{Hilb}(E,H)$ *is smooth in a neighbourhood of the point F.*

But now the exact sequence

$$0 \longrightarrow F' \longrightarrow E \longrightarrow F \longrightarrow 0$$

and the fact that $\mathrm{Ext}^q(E,F) = 0$ for $q > 0$ give an isomorphism $\mathrm{Ext}^1(F',F) \simeq \mathrm{Ext}^2(F,F)$. Hence, the Hilbert scheme $\mathbf{Hilb}(E,H)$ is smooth in a neighbourhood of the point F since $\mathrm{Ext}^2(F,F) = 0$. □

Example.
If X is the projective plane then the open subset $\mathbf{Hilb}^{ss}(E,H)$ of the Hilbert scheme is smooth. Indeed, it suffices to check that if F is a semi-stable sheaf then we have $\mathrm{Ext}^2(F,F) = 0$. But this follows from the definition of semi-stability and from Serre duality for Ext groups (cf. [15]):

Lemma 14.5.3 *Let F and G be two coherent algebraic sheaves on a smooth projective variety X of dimension n. Let ω_X be the dualizing sheaf. Then we have a natural pairing*

$$\mathrm{Ext}^p(F,G) \times \mathrm{Ext}^{n-p}(G, F \otimes \omega_X) \longrightarrow \mathbb{C}$$

which makes the vector spaces dual to each other.

In the case of the projective plane the dualizing sheaf is isomorphic to $\mathcal{O}(-3)$. If F is semi-stable then $\mathrm{Hom}(F, F(-3)) = 0$ and it follows that $\mathrm{Ext}^2(F,F) = 0$. This implies that the open subset $\mathbf{Hilb}^{ss}(E,H)$ of the Hilbert scheme, assuming it is non-empty, is smooth of dimension

$$r^2(2\Delta - 1) + (H(m))^2.$$

Corollary 14.5.4 *Let $\mathbf{M} = \mathbf{M}_{\mathbf{P}_2}(r, c_1, c_2)$ be the moduli space of semi-stable sheaves on the projective plane. Let χ denote the Euler-Poincaré characteristic and we assume that r, c_1 and χ are relatively coprime. Then $\mathbf{M}$ is a smooth projective variety of dimension $r^2(2\Delta - 1) + 1$.*

Indeed, if r, c_1 and χ are relatively coprime then all semi-stable sheaves are stable.

In general, the moduli space $\mathbf{M}$ is singular at the non-stable points. However, in some particular cases this is not true: for example, the moduli space $\mathbf{M}_{\mathbf{P}_2}(2,0,2)$ is isomorphic to $\mathbf{P}_5$ but it does contain some non-stable points.

14.6 The universal sheaf

Definition 14.6.1 *Let X be a polarized irreducible smooth algebraic surface. Let $f : S \to \mathbf{M}_X(H)$ be a map of algebraic varieties. A flat family F of semi-stable sheaves on X, parametrized by an algebraic variety, is called a* universal sheaf *(relative to f) if the associated moduli map f_F is equal to f.*

If $S = \mathbf{M}_X(H)$ and $f = \mathrm{id}$ then we simply say that F is a universal sheaf.

Such universal sheaves do not always exist. On the projective plane we have the following result.

Theorem 14.6.2 *Consider the moduli space $\mathbf{M} = \mathbf{M}_{\mathbf{P}_2}(r, c_1, c_2)$ on the projective plane and let χ denote the Euler-Poincaré characteristic.*

(i) If (r, c_1, χ) are relatively coprime then there exists a universal sheaf on $\mathbf{M}$.

(ii) Suppose that (r, c_1, χ) are not relatively coprime and consider a non-empty open subset U of $\mathbf{M}$ which meets the open set of stable points. Then there does not exist a universal sheaf parametrized by U.

The proof of this result calls on the same techniques that we will develop in chapter 18 to compute the Picard group and on the existence theorem which allows us to describe the Chern classes for which the moduli space is non-empty. Part (i) follows from a general result due to Maruyama and part (ii) is due to Drézet and Narasimhan [16].

14.7 A variant of the construction on the projective plane

Set $V^* = H^0(\mathcal{O}_{\mathbf{P}_2}(1))$. Suppose that the sheaves we consider are normalized, that is, such that $-r < c_1 \leq 0$, and we set

$$n = c_2 - \frac{1}{2}c_1(c_1 + 1),$$

where r is the rank and c_1 and c_2 are the Chern classes. We then have $n + c_1 \geq 0$ and $n \geq r + c_1$, and we know by Lemma 13.1.3, that a normalized semi-stable sheaf F on the projective plane can be obtained as the cohomology of a Beilinson monad. Let $H_0 = H^1(F(-1))$. In the notation of the lemma, this monad allows us to identify the vector space $H_1 = H^1(F(-2))$ with a vector sub-space of $H_0 \otimes V$ of dimension $n + c_1$

and the vector space $H_2 = H^1(F)$ with a quotient $H_0 \otimes V^*$ of dimension $n - (c_1 + r)$.

Conversely, fix a dimension n vector space H_0. To each point (K, L) of the variety

$$\mathfrak{P} = \mathbf{Grass}_{n+c_1}(H_0 \otimes V) \times \mathbf{Grass}^{n-(r+c_1)}(H_0 \otimes V^*),$$

where the first factor is the Grassmannian of sub-spaces of dimension $n + c_1$ and the second is the Grassmannian of quotient spaces of dimension $n - (r + c_1)$, we associate the maps

$$K \otimes \mathcal{O}_{\mathbf{P}_2}(-1) \xrightarrow{u} H_0 \otimes Q^* \xrightarrow{v} L \otimes \mathcal{O}_{\mathbf{P}_2}. \tag{14.1}$$

Let $\mathfrak{M}$ be the closed sub-scheme of $\mathfrak{P}$ given by the equation $vu = 0$.

The Picard group of $\mathfrak{P}$ is canonically isomorphic to $\mathbb{Z}^2$; the generators are induced from the bundle $\mathcal{O}(1)$ when we embed each of the Grassmannians into projective space using the Plücker embedding. For m sufficiently large, consider the polarization of $\mathfrak{P}$ given by the invertible bundle $\mathcal{O}((r + c_1)m - n, -c_1 m + n)$ and the semi-stable points under the action of the group $\mathrm{SL}(H_0)$. The following theorem gives another description of the moduli space $\mathbf{M}_{\mathbf{P}_2}(r, c_1, c_2)$ of S-equivalence classes of semi-stable sheaves of rank r and Chern classes c_1 and c_2 on the projective plane.

Theorem 14.7.1 [40] *Suppose that $\mathfrak{M}$ is equipped with the polarization $\mathcal{O}((r + c_1)m - n, -c_1 m + n)$ with m sufficiently large.*

(i) If (K, L) is a point of $\mathfrak{M}$, semi-stable under the action of $\mathrm{SL}(H_0)$ then the complex (14.1) above is a monad and its cohomology sheaf F is semi-stable of rank r and Chern classes c_1 and c_2.

(ii) Let $\mathfrak{M}^{ss}$ be the open subset of $\mathfrak{M}$ of semi-stable points. The universal family of sheaves F we thus obtain, parametrized by the open set $\mathfrak{M}^{ss}$, gives rise to an isomorphism

$$\mathfrak{M}^{ss}/\mathrm{SL}(H_0) \longrightarrow \mathbf{M}_{\mathbf{P}_2}(r, c_1, c_2),$$

where the first term is the Mumford quotient.

In this description, the variety $\mathfrak{M}$ replaces the Hilbert scheme. Its structure is very much simpler and allows us, for example, to understand immediately what the generators of the Chow ring are (at least when r, c_1 and c_2 are pairwise coprime) which is not obvious in the general description. It should also be useful for finding the relations between the generators [19].

15

Differential Study of the Shatz Stratification

The point of this chapter is to find the conditions which allow us to check that, in the family of torsion-free coherent sheaves on the projective plane, parametrized by a smooth variety S, the unstable sheaves form a strict closed subset. For curves, the proof given in chapter 8 shows that it suffices to consider complete families, that is, those whose Kodaira-Spencer infinitesimal deformation map surjects: this is what allows us to obtain the existence theorems for semi-stable bundles. In fact, under these conditions, the Harder-Narasimhan strata – also called the Shatz strata – are smooth and we can easily compute their codimension (cf. Corollary 15.4.3). To prove this result we need to start by extending Proposition 8.6.4 to the relative flag scheme. On surfaces, the situation is more delicate, and even for complete families, the Harder-Narasimhan strata need not be smooth. The smoothness of these strata for the projective plane is therefore obtained under a supplementary hypothesis (cf. 15.4.5). On the projective plane, in a complete family of "prioritary" coherent sheaves F_s, that is, satisfying the condition

$$\mathrm{Ext}^2(F_s, F_s(-1)) = 0,$$

this hypothesis is satisfied outside a closed set of codimension at least 2. Our principal reference for this chapter is [15]. The notion of prioritary sheaf is due to Hirschowitz and Laszlo [24].

15.1 The Kodaira-Spencer map

Let S be a smooth variety and X a smooth projective variety. Consider a coherent algebraic sheaf E on $Y = S \times X$, which is S-flat. For all $s \in S$ we let E_s denote the sheaf induced on the fibre $Y_s \simeq X$ of $S \times X$ over the point s. Let $\mathfrak{m}_s$ be the ideal of s. Consider the infinitesimal neighbourhood of order 1 of the fibre Y_s in Y denoted by Y_{s^2} given by the ideal $\mathfrak{m}_s^2$. The exact sequence of sheaves over Y_{s^2}

$$0 \longrightarrow T_s^* S \otimes E_s \longrightarrow E/\mathfrak{m}_s^2 E \longrightarrow E_s \longrightarrow 0$$

can be viewed as an exact sequence of sheaves on Y_s via the projection $Y_{s^2} \to Y_s$. This extension gives rise to an element $\omega \in \mathrm{Ext}^1(E_s, T_s^* S \otimes E_s)$; in other words, a linear map

$$\omega : T_s S \longrightarrow \mathrm{Ext}^1(E_s, E_s)$$

which is called the *Kodaira-Spencer infinitesimal deformation map* at the point $s \in S$.

Definition 15.1.1 *We say that the sheaf defines a* complete *family parametrized by S if at each point of S the Kodaira-Spencer infinitesimal deformation map surjects.*

15.2 The relative Hilbert scheme

Consider the relative Hilbert scheme $\mathbf{Hilb}(E/S, H) \xrightarrow{\pi} S$ of pairs (s, F) consisting of a point $s \in S$ and a coherent quotient sheaf E_s/F with fixed Hilbert polynomial H. Then this scheme represents the functor which associates to an algebraic variety $f : S' \to S$ over S, the set of coherent algebraic quotient sheaves $E_{S'}/F$ which are flat over S' and such that for all $s' \in S'$ the Hilbert polynomial of the induced quotient is H. Here, the sheaf $E_{S'}$ denotes the coherent sheaf on $S' \times X$ obtained from E under the change of base $f : S' \to S$.

Proposition 15.2.1 *Let $t = (s, F)$ be a point of the Hilbert scheme $\mathbf{Hilb}(E/S, H) \to S$. We have an exact sequence*

$$0 \to \mathrm{Hom}(F, E_s/F) \to T_t\,\mathbf{Hilb}(E/S, H) \xrightarrow{T_t \pi} T_s S \xrightarrow{\omega_+} \mathrm{Ext}^1(F, E_s/F)$$

in which the map ω_+ is the composite linear map

$$\begin{array}{ccc} T_s S & \xrightarrow{\quad \omega \quad} & \mathrm{Ext}^1(E_s, E_s) \\ & \searrow^{\omega_+} & \Big\downarrow \text{can} \\ & & \mathrm{Ext}^1(F, E_s/F) \end{array}$$

of the Kodaira-Spencer infinitesimal deformation map at the point s *and the canonical map associated to the inclusion* $F \hookrightarrow E_s$ *and the projection* $E_s \to E_s/F$.

We have already seen the same result for curves (Proposition 8.6.4). The proof is similar. The first part of the exact sequence gives rise to the fact that the fibre of π over the point s is just the absolute Hilbert scheme $\mathbf{Hilb}(E_s, H)$ and to the usual description of the Zariski tangent space to such a Hilbert scheme. For the exactness at the term T_sS we use the double point $D = \mathrm{Spec}(\mathbb{C}[\epsilon]/\epsilon^2)$ and interpret a tangent vector to a variety Z at a point $z \in Z$ as a map $u : D \to Z$ such that $u(0) = z$, where 0 denotes the closed point of D.

15.3 The cohomology spaces $\mathrm{Ext}^q_+(E, E)$ of a filtered sheaf

We need to generalize this result to flag varieties. We shall first define the space which plays the role of the vector space $\mathrm{Ext}^q(F, E_s/F)$ in the case of a flag. For the following we fix an integer $\ell \geq 2$. Consider two complexes of algebraic sheaves on a variety X which are equipped with finite filtrations

$$0 \subset F_1 \subset \cdots \subset F_\ell = K$$

and

$$0 \subset F'_1 \subset \cdots \subset F'_\ell = K'$$

of the same length. Suppose that the associated grading $\mathrm{gr}_i(K')$ is a complex of injective algebraic sheaves. Consider the homomorphism complex $\mathrm{Hom}^\bullet(K, K')$. We have an exact sequence of complexes

$$0 \to \mathrm{Hom}^\bullet_-(K, K') \to \mathrm{Hom}^\bullet(K, K') \to \mathrm{Hom}^\bullet_+(K, K') \to 0,$$

where the sub-complex $\mathrm{Hom}^\bullet_-(K, K')$ consists of homomorphisms $f \in \mathrm{Hom}^p(K, K')$ which respect the filtration, that is, such that $f(F_i) \subset F'_i$. The complex $\mathrm{Hom}^\bullet_+(K, K')$ is defined to be the quotient of $\mathrm{Hom}^\bullet(K, K')$ by this complex.

Definition 15.3.1 *We set*

$$\mathrm{Ext}^q_-(K, K') = H^q(\mathrm{Hom}^\bullet_-(K, K'))$$

and

$$\mathrm{Ext}^q_+(K, K') = H^q(\mathrm{Hom}^\bullet_+(K, K')).$$

In these definitions, the symbol H^q denotes cohomology vector spaces. So we have an exact sequence

$$\cdots \to \mathrm{Ext}^q_-(K,K') \to \mathrm{Ext}^q(K,K') \to \mathrm{Ext}^q_+(K,K') \to \mathrm{Ext}^{q+1}_-(K,K') \to \cdots$$

Moreover, we can define canonical increasing filtrations on each of the homomorphism complexes $\mathrm{Hom}^\bullet_-(K,K')$ and $\mathrm{Hom}^\bullet_+(K,K')$; for example, for $\mathrm{Hom}_+(K,K')$, the grading is given by

$$\mathrm{gr}_p \mathrm{Hom}^\bullet_+(K,K') = \begin{cases} \bigoplus_i \mathrm{Hom}^\bullet(\mathrm{gr}_i(K), \mathrm{gr}_{i+p}(K')) & \text{if } p > 0, \\ 0 & \text{if } p \leq 0. \end{cases}$$

We can then write down the spectral sequence of a filtered complex in cohomology. This gives us a spectral sequence whose E_1 term is given by

$$E_1^{p,q} = \begin{cases} \bigoplus_i \mathrm{Ext}^{p+q}(\mathrm{gr}_i(K), \mathrm{gr}_{i-p}(K')) & \text{if } p < 0, \\ 0 & \text{if } p \geq 0, \end{cases}$$

and which abuts on $\mathrm{Ext}^{p+q}_+(K,K')$ in degree $p+q$.

Now consider two algebraic sheaves on X together with finite increasing filtrations as above

$$0 \subset F_1 \subset \cdots \subset F_\ell = E$$

and

$$0 \subset F'_1 \subset \cdots \subset F'_\ell = E'$$

of the same length ℓ. We can find a filtered injective resolution of E': it is a filtered complex K' as above whose grading consists of injective sheaves together with a map $E' \to K'$ compatible with the filtrations and such that the induced maps on the gradings

$$\mathrm{gr}_i(E') \longrightarrow \mathrm{gr}_i(K')$$

are quasi-isomorphisms for $i = 1, \ldots, \ell$. Then we define the vector spaces $\mathrm{Ext}^q_-(E,E')$ and $\mathrm{Ext}^q_+(E,E')$ by setting

$$\mathrm{Ext}^q_-(E,E') = \mathrm{Ext}^q_-(E,K') \text{ and } \mathrm{Ext}^q_+(E,E') = \mathrm{Ext}^q_+(E,K').$$

One can show that these do not depend on the resolution. All of the statements above concerning filtered complexes extend to filtered coherent algebraic sheaves. So we have, for example, a spectral sequence whose E_1 term is given by

$$E_1^{p,q} = \begin{cases} \bigoplus_i \mathrm{Ext}^{p+q}(\mathrm{gr}_i(E), \mathrm{gr}_{i-p}(E)) & \text{if } p < 0, \\ 0 & \text{if } p \geq 0 \end{cases}$$

and which abuts on $\mathrm{Ext}^{p+q}_+(E,E)$ in degree $p+q$. Similarly, we have a spectral sequence whose E_1 term is given by

$$E_1^{p,q} = \begin{cases} 0 & \text{if } p<0, \\ \bigoplus_i \mathrm{Ext}^{p+q}(\mathrm{gr}_i(E), \mathrm{gr}_{i-p}(E)) & \text{if } p \geq 0, \end{cases}$$

and which abuts on $\mathrm{Ext}^{p+q}_-(E,E)$ in degree $p+q$.

Proposition 15.3.2 *Let E be a locally-free sheaf on a smooth algebraic projective curve X of genus g equipped with its Harder-Narasimhan filtration. We let r_i and μ_i denote the ranks and slopes of the Harder-Narasimhan grading. Then*

$$\mathrm{Ext}^0_+(E,E) = 0$$

and

$$\dim \mathrm{Ext}^1_+(E,E) = \sum_{i<j} r_i r_j (\mu_i - \mu_j + g - 1).$$

Proof We have $\mathrm{Hom}(\mathrm{gr}_i(E), \mathrm{gr}_j(E)) = 0$ for $i<j$. It follows from the first of the spectral sequences above that $E_1^{p,q} = 0$ if $p+q=0$. So the spectral sequence degenerates and

$$\dim \mathrm{Ext}^1_+(E,E) = \sum_{i<j} \dim \mathrm{Ext}^1(\mathrm{gr}_i(E), \mathrm{gr}_j(E)).$$

The right hand side can be computed from the Riemann-Roch formula, from which we obtain the result. □

Proposition 15.3.3 *Let E be a torsion-free coherent sheaf of rank r on the projective plane $X = \mathbf{P}_2$ together with its Harder-Narasimhan filtration. We let (r_i, μ_i, Δ_i) denote the rank, slope and discriminant, respectively, of $\mathrm{gr}_i(E)$. Suppose that $\mu_{\max}(E) - \mu_{\min}(E) < 3$. Then*

$$\mathrm{Ext}^0_+(E,E) = \mathrm{Ext}^2_+(E,E) = 0.$$

$$\dim \mathrm{Ext}^1_+(E,E) = \sum_{i<j} r_i r_j (\Delta_i + \Delta_j - P(\mu_j - \mu_i)),$$

where P is the Hilbert polynomial of the trivial bundle of rank 1 on $\mathbf{P}_2$.

Proof Again we use the first of the spectral sequences above. Clearly, if $i<j$ then we have $\mathrm{Hom}(\mathrm{gr}_i(E), \mathrm{gr}_j(E)) = 0$ and so $E_1^{p,q} = 0$ if $p+q=0$. On the other hand, by the Serre Duality Theorem for Ext, 14.5.3, the hypothesis of the proposition implies that $\mathrm{Ext}^2(\mathrm{gr}_i(E), \mathrm{gr}_j(E)) = 0$ for

$i < j$. This shows that the $E_1^{p,q}$ terms in the spectral sequence are zero if $p+q \neq 1$. Hence, $\dim \operatorname{Ext}^1_+(E,E) = \sum_{i<j} \dim \operatorname{Ext}^1(\operatorname{gr}_i(E), \operatorname{gr}_j(E))$. The right hand side can be computed from the Riemann-Roch Theorem:

Lemma 15.3.4 *Let F' and F'' be two coherent algebraic sheaves on the projective plane of ranks r' and r'', slopes μ' and μ'' and discriminants Δ' and Δ'', respectively. Set*

$$\chi(F', F'') = \sum_i (-1)^i \dim_{\mathbf{C}} \operatorname{Ext}^i(F', F'').$$

Then

$$\chi(F', F'') = r'r''(P(\mu'' - \mu') - \Delta' - \Delta''),$$

where P denotes the Hilbert polynomial of the trivial rank 1 bundle on $\mathbf{P}_2$.

Proof As we have seen in chapter 12, the formula holds if F' is locally-free: it follows from the Riemann-Roch formula applied to the local Hom sheaf $\mathcal{H}om(F', F'')$. When F' is not locally-free we can reduce to this situation using the obvious additive properties and choosing a finite left resolution of F' by locally-free sheaves. This completes the proof of the lemma and the proposition. □

15.4 The scheme of relative flags

Now consider a projective algebraic variety $Y \to S$ over an algebraic variety S and a coherent algebraic sheaf E flat over S. Let P_i be some polynomials. For any algebraic variety $S' \to S$, consider the functor which associates to S' the set of flags

$$0 \subset F_1 \subset \cdots \subset F_\ell = E_{S'}$$

of coherent sub-sheaves satisfying the following conditions:

(i) the associated grading gr_i is an S'-flat sheaf;
(ii) for all $s \in S'$, the sheaf $\operatorname{gr}_{i,s}$ induced on the fibre Y_s has Hilbert polynomial P_i.

This functor can be represented by a projective algebraic variety over S

$$\mathbf{Flag}(E/S; P_1, \ldots, P_\ell) \xrightarrow{\pi} S.$$

The fibre over a point s is the absolute flag scheme of E_s whose associated grading gr_i has Hilbert polynomial P_i. The vertical tangent space at a point corresponding to a filtration F of E_s can be identified with the vector space $\mathrm{Ext}^0_+(E_s, E_s)$ associated to that filtration.

In the following proposition we use the hypothesis $Y = S \times X$, where X is a projective variety so that the Kodaira-Spencer infinitesimal deformation space makes sense. It generalizes Proposition 15.2.1 which corresponds to the case $\ell = 2$.

Proposition 15.4.1 *Let $t = (s, (F_1, \ldots, F_\ell))$ be a point of the Hilbert scheme of flags $\mathbf{Flag}(E/S; P_1, \ldots, P_\ell) = \mathbf{Flag}$. Equip the sheaf E_s with this filtration. Then we have a natural exact sequence*

$$0 \longrightarrow \mathrm{Ext}^0_+(E_s, E_s) \longrightarrow T_t\,\mathbf{Flag} \xrightarrow{T_t\pi} T_sS \xrightarrow{\omega_+} \mathrm{Ext}^1_+(E_s, E_s),$$

where the map ω_+ is the composite map

$$\begin{array}{ccc} T_sS & \xrightarrow{\quad\omega\quad} & \mathrm{Ext}^1(E_s, E_s) \\ & {\scriptstyle\omega_+}\searrow & \Big\downarrow {\scriptstyle\mathrm{can}} \\ & & \mathrm{Ext}^1_+(E_s, E_s) \end{array}$$

We shall apply this exact sequence to calculate the codimension of the Shatz strata. First of all we give a criterion which allows us to check whether certain flag schemes that we have introduced are smooth. This allows us to find conditions under which the Shatz strata are smooth and to determine the exact dimension of the strata. It is possible to avoid using this criterion in what follows but then we only obtain an upper bound for the dimension of the strata. This criterion is proved in [15]. It rests on the usual criterion of Grothendieck that we have already used, generalized to flag schemes, which allowed us to check that certain Hilbert schemes were smooth.

Proposition 15.4.2 *We use the notation of Proposition 15.4.1. Suppose that S is smooth and that the following conditions are satisfied:*

(i) $\mathrm{Ext}^2(E_s, E_s) = 0$;

(ii) the map ω_+ associated to a flag $(F_1, \ldots, F_\ell)$ of E_ℓ surjects.

Then the scheme $\mathbf{Flag}(E/S; P_1, \ldots, P_\ell)$ is smooth in a neighbourhood of $(s, (F_1, \ldots, F_\ell))$.

In the case of curves we obtain the following corollary whose idea is essentially due to Atiyah and Bott [3].

Corollary 15.4.3 *Consider a projective curve X and a complete family of vector bundles E_s parametrized by a smooth variety S. For integers ℓ, $r_i > 0$ and d_i set*

$$\mu_i = \frac{d_i}{r_i}.$$

The points $s \in S$ such that the Harder-Narasimhan filtration (if it exists) has length ℓ and such that the Harder-Narasimhan grading $\mathrm{gr}_i(E_s)$ of E_s has rank r_i and degree d_i for $i = 1, \ldots, \ell$, form a locally closed smooth sub-variety of codimension

$$\sum_{i<j} r_i r_j (\mu_i - \mu_j + g - 1).$$

Proof Let $\mathfrak{p}$ be the polygonal line defined by the integers r_i and d_i which we can assume to be concave. We have already seen that the points $s \in S$ such that the polygonal line $\mathfrak{p}_F$ associated to the Harder-Narasimhan filtration F of E_s is $> \mathfrak{p}$ form a closed set $S_{\mathfrak{p}}$. Let S' denote the complementary open set. Then the image of the induced map

$$\pi : \mathbf{Flag}(E/S'; P_1, \ldots, P_\ell) \longrightarrow S'$$

is exactly $S_{\mathfrak{p}}$. The hypotheses of Proposition 15.4.2 are satisfied. This flag variety is smooth and the map π is proper, injective and unramified, and so $S_{\mathfrak{p}}$ is itself a closed smooth sub-variety of S'; its normal space at a point can be identified from the exact sequence above to the vector space $\mathrm{Ext}^1_+(E_s, E_s)$ associated to the Harder-Narasimhan filtration of E_s. □

Example.
A vector bundle F on $X = \mathbf{P}_1$ is called *rigid* (respectively *quasi-rigid*) if $\mu_{\max}(F) - \mu_{\min}(F) \leq 1$ (respectively ≤ 2). This also means that $\mathrm{Ext}^1(F, F) = 0$. If we consider a complete family of vector bundles $E_s \to X$ of rank r and degree d parametrized by a smooth variety S then the above corollary leads, in particular, to

(1) the points $s \in S$ such that F_s are not rigid form a closed subset of codimension 1 if $d \equiv 0 \bmod r$ and of codimension at least 2 if $d \not\equiv 0 \bmod r$,

(2) the points $s \in S$ such that F_s are not quasi-rigid form a closed set of codimension at least 2.

Corollary 15.4.4 *Consider a complete family of torsion-free coherent sheaves F_s on the projective plane parametrized by a smooth algebraic variety S and satisfying the following condition:*

$$\text{(P)} \qquad \textit{For all } s \in S, \ \mathrm{Ext}^2(F_s, F_s(-1)) = 0.$$

Then the set of points $s \in S$ such that $\mu_{\max}(F_s) - \mu_{\min}(F_s) > 2$ forms a closed subset of S of codimension at least 2.

The coherent sheaves which satisfy condition P are called *prioritary* sheaves by Hirschowitz and Laszlo [24].

Proof Let d be a line of $\mathbf{P}_2$. Consider the open subset $S_d \subset S$ corresponding to sheaves F_s such that F_s is non-singular on d. Then we obtain a family $F_s|_d$ of bundles on the line d and this family remains complete: indeed, the Kodaira-Spencer infinitesimal deformation map is just the composite linear map

$$T_sS \longrightarrow \mathrm{Ext}^1(F_s, F_s) \longrightarrow \mathrm{Ext}^1_d(F_s|_d, F_s|_d),$$

where the first map surjects because the family is complete and the second is the restriction map. To show that this restriction map surjects it suffices to write down that derived long exact sequence of the exact sequence

$$0 \longrightarrow F_s(-1) \longrightarrow F \longrightarrow F_s|_d \longrightarrow 0$$

by applying the functor $\mathrm{Hom}(F_s, -)$. It suffices to note that we have $\mathrm{Ext}^2(F_s, F_s(-1)) = 0$ from hypothesis (P). We can then apply the conclusions of the example above to this family. Since the open sets S_d obviously cover S, we obtain, in particular, that the points $s \in S$ such that $\mu_{\max}(F_s) - \mu_{\min}(F_s) > 2$ form a closed set of codimension at least 2. □

Example.
In particular, hypothesis (P) is satisfied for a family of semi-stable sheaves because of the Serre Duality Theorem.

Corollary 15.4.5 *Consider a complete family of torsion-free coherent algebraic sheaves F_s on the projective plane, parametrized by a smooth variety S. We make the following additional hypothesis:*

$$\text{(R)} \qquad \mu_{\max}(F_s) - \mu_{\min}(F_s) < 3.$$

Let $\ell \geq 2$ be an integer and $P_1, \ldots, P_\ell$ polynomials. Consider the set $S(P_1, \ldots, P_\ell)$ of points $s \in S$ such that the Harder-Narasimhan filtration has length ℓ and the Harder-Narasimhan grading $\mathrm{gr}_i(F_s)$ of F_s has Hilbert polynomial P_i for $i = 1, \ldots, \ell$. This means that the rank r_i, slope μ_i and discriminant Δ_i of gr_i are determined by P_i. Then, if non-empty, $S(P_1, \ldots, P_\ell)$ is a smooth locally closed sub-variety of codimension

$$\sum_{i<j} r_i r_j (\Delta_i + \Delta_j - P(\mu_j - \mu_i)).$$

Proof The proof is almost identical to that which we have given for curves. The only difference is that we must check that $\mathrm{Ext}^2(F_s, F_s) = 0$ and that the map ω_+ which appears in the exact sequence of 15.4.1 surjects. But, taking Proposition 15.3.3 into account, it suffices to note that the hypotheses imply that $\mathrm{Ext}^2_-(F_s, F_s) = 0$ for the Harder-Narasimhan filtration. For this we use the spectral sequence whose $E_1^{p,q}$ term is given by

$$E_1^{p,q} = \begin{cases} \bigoplus_i \mathrm{Ext}^{p+q}(\mathrm{gr}_i(F_s), \mathrm{gr}_{i-p}(F_s)) & \text{if } p \geq 0, \\ 0 & \text{if } p < 0, \end{cases}$$

and which abuts on $\mathrm{Ext}_-^{p+q}(F_s, F_s)$ in degree $p + q$. Then it suffices to check that, for $j \leq i$,

$$\mathrm{Ext}^2(\mathrm{gr}_i(F_s), \mathrm{gr}_j(F_s)) = 0$$

which follows trivially from the Serre Duality Theorem. Hypothesis (R) allows us to apply Proposition 15.3.3 which enables us to compute the dimension of the normal space. □

Corollary 15.4.6 *Consider a complete family F_s of semi-stable sheaves with fixed Hilbert polynomials on the projective plane, parametrized by a smooth algebraic variety S of dimension n. Let Δ be the discriminant of these sheaves.*

If $\Delta > 1/2$ then the points $s \in S$ such that F_s are stable forms an open subset whose complement has codimension at least 1.

Proof We apply Proposition 15.4.1 to the flag corresponding to a Jordan-Hölder filtration of $E = F_s$. We still have $\mathrm{Ext}^2(\mathrm{gr}_i(E), \mathrm{gr}_j(E)) = 0$ for all (i, j) for such a filtration. Thus, the map ω_+ surjects and we have

$$\dim \mathrm{Ext}^1_+(E, E) \geq -\sum_{i<j} \chi(\mathrm{gr}_i(E), \mathrm{gr}_j(E)) = (2\Delta - 1) \sum_{i<j} r_i r_j.$$

The hypothesis $2\Delta > 1$ implies that the tangent map ω_+ has rank less than n, so the image of π is closed and has codimension less than n. The result now follows. □

16

The Conditions for Existence

In this chapter we shall answer the question: for which Chern classes do there exist semi-stable sheaves on the projective plane? Bogomolov's Theorem gives us a necessary condition: we must have $\Delta \geq 0$. But this condition is not sufficient to ensure that such sheaves exist: we have already seen an example where $r = 2$ for which there are no semi-stable sheaves with Chern classes $c_1 = 0$ and $c_2 = 1$. The description of the existence conditions necessitates the introduction of exceptional bundles, for more details, the reader can consult the treatment given in [15].

16.1 Exceptional bundles

Proposition 16.1.1 *Let F be a stable sheaf on the projective plane with discriminant Δ. The following are equivalent:*

(i) The sheaf F is rigid, *that is,*

$$\operatorname{Ext}^1(F, F) = 0;$$

(ii) $\Delta < 1/2$;

(iii) $\chi(F, F) > 0$.

Proof Serre duality shows that $\operatorname{Ext}^2(F, F) = 0$. A stable bundle has no endomorphisms other than homotheties. Therefore,

$$\chi(F, F) = 1 - \dim \operatorname{Ext}^1(F, F) = r^2(1 - 2\Delta).$$

This leads to the required equivalence. □

Definition 16.1.2 *A stable (respectively semi-stable) sheaf F on the projective plane is called an* exceptional *(respectively* semi-exceptional*) sheaf if it satisfies condition (i) of Proposition 16.1.1.*

Examples.

(i) Any invertible bundle is exceptional.

(ii) The kernel of the evaluation map

$$H^0(\mathcal{O}_{\mathbf{P}_2}(i)) \otimes \mathcal{O}_{\mathbf{P}_2} \longrightarrow \mathcal{O}_{\mathbf{P}_2}(i)$$

is an exceptional bundle if and only if $i = 1$ or $i = 2$. For $i = 1$ it is the bundle $Q^* \simeq T(-2)$, of rank 2 and Chern classes $(-1, 2)$. For $i = 2$ it is a rank 5 bundle with Chern classes $(-2, 4)$.

(iii) If F is an exceptional bundle then the bundle $F(i)$ is also exceptional.

Proposition 16.1.3 *For all $\alpha \in \mathbb{Q}$, there exists at most one exceptional sheaf of slope α. This sheaf is locally-free and μ-stable. Its rank r is the least positive integer such that $r\alpha \in \mathbb{Z}$ and its discriminant Δ is given by the formula*

$$\Delta = \frac{1}{2}\left(1 - \frac{1}{r^2}\right).$$

Proof Let (r, c_1, c_2) be the rank and Chern classes of an exceptional sheaf F. Then we have

$$\chi(F, F) = 1 = r^2 - 2rc_2 + (r-1)c_1^2 = r^2(1 - 2\Delta)$$

and it follows that the integers r and c_1 are coprime. Hence, r and c_1 are determined by the slope $\alpha = c_1/r$, and the discriminant Δ can be computed from the rank. If F' is another exceptional sheaf of the same slope α then F and F' have the same Chern classes and so $\chi(F, F') = 1$. Then we have $\mathrm{Ext}^2(F, F') = 0$ and so $\mathrm{Hom}(F, F') \neq 0$ and the stability property implies that $F \simeq F'$. In particular, if $g \in \mathrm{Aut}(\mathbf{P}_2)$, the sheaf $g^*(F)$ is still an exceptional sheaf of the same slope and so is isomorphic to F. Thus, the sheaf F is invariant under the action of $\mathrm{Aut}(\mathbf{P}_2)$ and so is locally-free. □

Corollary 16.1.4 *The dual of an exceptional bundle is an exceptional bundle.*

Note that, if F is an exceptional bundle then there are no other semi-stable or stable sheaves of the same Chern classes as F. Hence:

Corollary 16.1.5 *If there exist an exceptional bundle of rank r and Chern classes c_1 and c_2 then the moduli space $\mathbf{M}_{\mathbf{P}_2}(r, c_1, c_2)$ is reduced to a point.*

Corollary 16.1.6 *A semi-exceptional sheaf of slope μ is a direct sum of exceptional bundles of slope μ.*

Proof Consider a Jordan-Hölder filtration of F. Then by Proposition 16.1.3, all of the gradings gr_i are isomorphic to an exceptional bundle E_μ. Indeed, if Δ is the discriminant of F then the grading gr_i has slope μ and discriminant Δ, and, by hypothesis, $\Delta < 1/2$. Then $\mathrm{Ext}^1(\mathrm{gr}_i, \mathrm{gr}_j) = 0$ and so the Jordan-Hölder filtration splits. $\square$

16.2 The Existence Theorem [15]

Let $\mathfrak{E}$ denote the set of rational numbers which are the slopes of exceptional bundles. This set is invariant under translation by integers and also under inversion in the origin. For $\alpha \in \mathfrak{E}$, we let r_α and Δ_α denote the rank and discriminant of the exceptional bundle determined by α.

Theorem 16.2.1 *Let r be an integer and μ and Δ two rational numbers. The following conditions are necessary and sufficient for the existence of a stable sheaf of rank r, slope μ and discriminant Δ on the projective plane:*

(i) $c_1 = r\mu \in \mathbb{Z}$, $\chi = r(P(\mu) - \Delta) \in \mathbb{Z}$;
(ii) for all $\alpha \in \mathfrak{E}$, such that $r_\alpha < r$ and $|\mu - \alpha| < 3$, we have

$$\text{(S)} \qquad \Delta + \Delta_\alpha \geq P(-|\mu - \alpha|).$$

Remark.
We shall see that conditions (i) and (ii) are actually satisfied for semi-stable sheaves of slope μ and discriminant $\Delta \geq 1/2$. Condition (ii) fails for semi-exceptional sheaves.

Proof : *The conditions are necessary:* Condition (i) just expresses the fact that the Chern class c_1 and the Euler-Poincaré characteristic are integers. Let F be a semi-stable sheaf of slope μ and discriminant Δ. We assume one of the following hypotheses:

(a) $\Delta \geq 1/2$;
(b) the sheaf F is stable.

Then for all $\alpha \in \mathfrak{E}$ such that $|\mu - \alpha| < 3$ we have

$$(\mathrm{S}^*) \qquad \Delta + \Delta_\alpha \geq P(-|\mu - \alpha|)$$

except if F is exceptional and $\alpha = \mu$. Indeed, let E_α denote an exceptional bundle of slope α. If $0 < \alpha - \mu < 3$ then the semi-stability condition and the Serre Duality Theorem imply

$$\mathrm{Hom}(E_\alpha, F) = 0 \qquad \text{and} \qquad \mathrm{Ext}^2(E_\alpha, F) = 0.$$

If $\mu = \alpha$ then we still have the same thing if F is not isomorphic to E_α because if $\mathrm{Hom}(E_\alpha, F) \neq 0$ then we would have $\Delta_\alpha \geq \Delta$ and then F would be the direct sum of bundles isomorphic to E_α, which is excluded by each of the hypotheses (a) and (b). It follows that $\chi(E_\alpha, F) \leq 0$. We can reason in the same way for $0 < \mu - \alpha < 3$ by considering $\chi(F, E_\alpha)$.

The conditions are sufficient: The result is obvious for $r = 1$. So suppose that $r \geq 2$. We shall actually construct a family of semi-stable bundles.

Construction of a large family of vector bundles

Hypothesis (i) ensures that the polynomial $H(i) = r(P(\mu + i) - \Delta)$ (which represents the Hilbert polynomial) has integer coefficients. The sequence $i \mapsto H(i) = r(P(\mu + i) - \Delta)$ is increasing for $\mu + i \geq -3/2$ and is non-positive for $-3 < \mu + i \leq 0$, by condition (ii). This implies that if we consider the first integer m such that $H(m) > 0$ then we have $H(m-1) \leq 0$ and $H(m-2) \leq 0$. We set $N_0 = H(m)$, $N_1 = -H(m-1)$ and $N_2 = -H(m-2)$ and consider the family of vector bundle maps

$$s : \mathcal{V} = \mathcal{O}_{\mathbf{P}_2}(-1)^{N_2} \longrightarrow \mathcal{W} = (Q^*)^{N_1} \oplus \mathcal{O}_{\mathbf{P}_2}^{N_1}.$$

This family is parametrized by the vector space $\mathrm{Hom}(\mathcal{V}, \mathcal{W})$. The difference of ranks between the vector bundles $\mathcal{V}$ and $\mathcal{W}$ is $H(m) - 2H(m-1) + H(m-2) = r$, and the homomorphism bundle $\mathcal{H}om(\mathcal{V}, \mathcal{W})$ is generated by its sections. By a variant of the Bertini-Sard Theorem, we see that the closed set of maps s which are not injective at at least one point of $\mathbf{P}_2$ has codimension $r-1$. Let S be the complementary open set. For $s \in S$, the cokernel sheaf G_s of s is a locally-free sheaf whose

Hilbert polynomial is $H(m+i)$. It follows that the family of locally-free sheaves $F_s = G_s(-m)$, parametrized by S, has Hilbert polynomial H. Hence, these vector bundles have rank r, slope μ and discriminant Δ. This family is obviously complete and satisfies condition (P) of the preceding chapter.

Construction of semi-stable vector bundles

We need to remove the bundles which are not semi-stable from the family above and to verify that the remaining open set is not empty. It follows from Corollary 15.4.4 that there exists a closed set $\Sigma \subset S$ of codimension at least 2 away from which we have $\mu_{\max}(F_s) - \mu_{\min}(F_s) \leq 2$. On the open set $S' = S \setminus \Sigma$, consider the Shatz strata corresponding to the sheaves $E = F_s$ whose Harder-Narasimhan filtration has length ℓ at least 2 and whose grading gr_i has Hilbert polynomial $r_i(P(\mu_i + m) - \Delta_i)$. We know from Corollary 15.4.5 that these strata have codimension

$$\sum_{i<j} r_i r_j \big(\Delta_i + \Delta_j - P(\mu_j - \mu_i)\big).$$

We shall show that this codimension is strictly positive if $\ell > 1$. Each of the terms in this sum is at least 0 and if this sum were zero, this would imply that $\chi(\mathrm{gr}_i, \mathrm{gr}_j) = 0$ for all $i < j$; in other words,

$$\Delta_i + \Delta_j = P(\mu_j - \mu_i).$$

In particular, this equality gives us $P(\mu_\ell - \mu_1) = \Delta_1 + \Delta_\ell \geq 0$. Recall that $0 \leq \mu_1 - \mu_\ell < 3$. Suppose that $\mu_1 - \mu_\ell > 0$. Then $\Delta_1 + \Delta_\ell < 1$ and so either Δ_1 or Δ_ℓ is less than $1/2$. If $\mu_1 = \mu_\ell$ then we still have $\Delta_1 < \Delta_\ell$ by the definition of the Harder-Narasimhan filtration. In either case one of the sheaves gr_1 and gr_ℓ is semi-exceptional.

Suppose, for example, that gr_1 is semi-exceptional. We would then have

$$\begin{aligned} \chi(\mathrm{gr}_1, E) &= \chi(\mathrm{gr}_1, \mathrm{gr}_1) + \sum_{j>1} \chi(\mathrm{gr}_1, \mathrm{gr}_j) \\ &= \chi(\mathrm{gr}_1, \mathrm{gr}_1) > 0. \end{aligned}$$

This means that $\Delta + \Delta_1 < P(\mu - \mu_1)$ which contradicts condition (ii). In the case where it is gr_ℓ which is semi-exceptional then we get another contradiction by studying $\chi(E, \mathrm{gr}_\ell)$.

Construction of stable bundles

If $\Delta < 1/2$ then we know that the sheaf F_s must be semi-exceptional. But then if F_s is not stable this contradicts condition (ii).

If $\Delta > 1/2$ then we can apply Corollary 15.4.6 to the complete family of semi-stable bundles $(F_s)_{s \in S''}$ constructed above which remain after we have removed the Shatz strata of positive codimension from S'. This corollary gives the existence of stable bundles.

The case $\Delta = 1/2$ is excluded by condition (S*): there do not exist semi-stable sheaves with discriminant 1/2. But to prove this result we need to study the arithmetic of $\mathfrak{E}$ in detail. □

16.3 Description of $\mathfrak{E}$

Definition 16.3.1 *Given a rational number α, the largest integer $r > 0$ such that $r\alpha$ is integral is denoted r_α. The rational numbers*

$$\Delta_\alpha = \frac{1}{2}\left(1 - \frac{1}{r_\alpha^2}\right) \qquad \textit{and} \qquad \chi_\alpha = r_\alpha(P(\alpha) - \Delta_\alpha)$$

are called the discriminant *and* characteristic *of α, respectively.*

Given two rational numbers $\alpha \in \mathfrak{E}$ and $\beta \in \mathfrak{E}$ such that $0 < \beta - \alpha < 3$, the discriminant Δ of a number $t \in \mathfrak{E}$ such that $\alpha < t < \beta$ satisfies the following conditions:

$$\Delta \geq P(\alpha - t) - \Delta_\alpha \qquad \text{and} \qquad \Delta \geq P(t - \beta) - \Delta_\beta.$$

Consider the intersection point of the parabolæ with equations

$$\Delta = P(\alpha - t) - \Delta_\alpha \qquad \text{and} \qquad \Delta = P(t - \beta) - \Delta_\beta$$

whose abscissa is given by

$$t = \frac{\alpha + \beta}{2} + \frac{\Delta_\beta - \Delta_\alpha}{3 + \alpha - \beta}.$$

This unique solution will be denoted by $\alpha.\beta$.

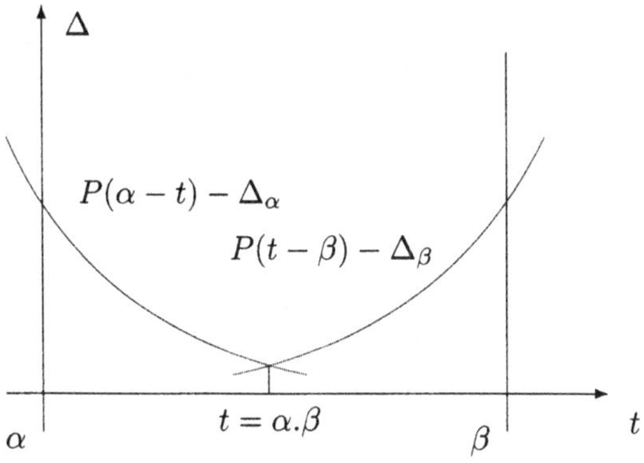

Definition 16.3.2 *A pair of rational numbers* (α, β) *will be called* admissible *if the following conditions are satisfied:*

$$(\text{A}) \qquad \begin{cases} 0 < \beta - \alpha < 3, \\ P(\alpha - \beta) = \Delta_\alpha + \Delta_\beta, \\ \chi_\alpha \in \mathbb{Z}, \quad \chi_\beta \in \mathbb{Z}, \end{cases}$$

For example $(0, 1)$ *and* $(0, 2)$ *are admissible pairs.*

Proposition 16.3.3 *Let* (α, β) *be an admissible pair of rational numbers. Then*

(i) $\alpha < \alpha.\beta < \beta$,

(ii) *the rational number* $\alpha.\beta$ *has rank* $r_{\alpha.\beta} = r_\alpha r_\beta (3 + \alpha - \beta)$,

(iii) *the pairs* $(\alpha, \alpha.\beta)$ *and* $(\alpha.\beta, \beta)$ *are admissible.*

This proposition, whose proof is an entirely elementary arithmetic computation, allows us to construct the set $\mathfrak{C}$. Consider the set $\mathfrak{D}$ of numbers of the form $p/2^q$, with $p \in \mathbb{N}$ and $q \in \mathbb{N}$, and construct an increasing map

$$\epsilon : \mathfrak{D} \longrightarrow \mathbb{Q}$$

by induction on q by setting

$$\begin{cases} \epsilon(n) = n & \text{for all } n \in \mathbb{Z}, \\ \epsilon\Big(\dfrac{2p+1}{2^{q+1}}\Big) = \epsilon\Big(\dfrac{p}{2^q}\Big).\epsilon\Big(\dfrac{p+1}{2^q}\Big). \end{cases}$$

The map ϵ is well defined and strictly increasing. It follows from the preceding proposition that the pair $\big(\epsilon(p/2^q), \epsilon((p+1)/2^q)\big)$ is admissible. In particular, $\epsilon(p/2^q)$ has integral characteristic and, if p is odd, then its rank is at least 2^q. One can further prove that ϵ commutes with translations by integers and by inversion in 0. So to describe ϵ it suffices to describe it on $[0, 1/2]$. For example, direct computation gives

$\frac{p}{2^q}$	0	$\frac{1}{8}$	$\frac{1}{4}$	$\frac{3}{8}$	$\frac{1}{2}$
$\epsilon\left(\frac{p}{2^q}\right)$	0	$\frac{5}{13}$	$\frac{2}{5}$	$\frac{12}{29}$	$\frac{1}{2}$

Theorem 16.3.4 *The set $\mathfrak{E}$ is the image of $\mathfrak{D}$ under the map ϵ defined above.*

The exceptional bundles are thus exactly determined up to isomorphism.

Proof (1) We establish the inclusion $\epsilon(\mathfrak{D}) \subset \mathfrak{E}$. We induct on q. Set $\alpha = \epsilon(p/2^q)$ and $\beta = \epsilon((p+1)/2^q)$. By the induction hypothesis, these elements are contained in $\mathfrak{E}$. We show that $\mu = \alpha.\beta$ is the slope of a stable bundle. It suffices to check conditions (i) and (ii) of the Existence Theorem 16.2.1. The integrality conditions have already been checked. It only remains to check the inequalities in (ii) which concern $\alpha' \in \mathfrak{E}$ such that $|\mu - \alpha'| < 3$ and such that $r_{\alpha'} < r_\mu$. They are clear for $\alpha' = \alpha$ or β. If $\alpha < \alpha' < \beta$ then we have $\Delta_{\alpha'} \geq \Delta_\mu$ from which we obtain $r_{\alpha'} \geq r_\mu$. We then only need to check the inequalities of (ii) for the $\alpha' \in \mathfrak{E}$ which are not contained in $[\alpha, \beta]$.

Suppose, for example, that $\alpha' < \alpha$. For $t \leq \beta$ we have the inequality

$$P(\alpha' - t) - \Delta_{\alpha'} \leq P(\alpha - t) - \Delta_\alpha.$$

Indeed, this inequality is satisfied at the point $t = \beta$ because of the fact that the pair (α, β) is admissible and from the necessary existence conditions (S*), and the general case follows from the fact that the function $t \mapsto P(t-\alpha) - P(t-\alpha')$ with derivative $\alpha' - \alpha$ is decreasing. Since (α, μ) is an admissible pair, we obtain the inequality $P(\alpha' - t) \leq \Delta_{\alpha'} - \Delta_\mu$ for $t = \mu$ which was what we required. The case when $\alpha' > \beta$ is analogous.

(2) The inclusion $\mathfrak{E} \subset \epsilon(\mathfrak{D})$. Let $\mu \in \mathfrak{E}$ and consider an integer q such that $r_\mu \leq 2^q$. Consider the set $\mathfrak{D}_q \subset \mathfrak{D}$ of rational numbers with denominator 2^q. The image of $\mathfrak{D}_q$ is locally finite and so we can find an integer $p \in \mathbb{Z}$ such that

$$\epsilon\left(\frac{p}{2^q}\right) \leq \mu < \epsilon\left(\frac{p+1}{2^q}\right).$$

Suppose that we have strict inequality. Let $\nu = \epsilon((2p+1)/(2^{q+1}))$. Because $\mu \in \mathfrak{E}$, the condition (S*) implies $\Delta_\mu \geq \Delta_\nu$ and then $r_\mu \geq r_\nu \geq$

2^{q+1}. This contradicts the definition of q. So $\mu = \epsilon(p/2^q)$ and $\mu \in \epsilon(\mathfrak{E})$.
□

Remark.
The exceptional bundles come with a natural action of the projective group $\mathrm{Aut}(\mathbf{P}_2)$. We can give a more explicit inductive construction along the lines of the way we constructed the exceptional bundle of slope $-2/5$, as the kernel of the canonical map

$$H^0(\mathcal{O}_{\mathbf{P}_2}(i)) \otimes \mathcal{O}_{\mathbf{P}_2} \longrightarrow \mathcal{O}_{\mathbf{P}_2}(i)$$

(cf. [12]).

16.4 The function $\mu \mapsto \delta(\mu)$

To write down the existence conditions of stable bundles of rank r and Chern classes c_1 and c_2, described in Theorem 16.2.1, we simply have a finite number of inequalities in terms of the slope μ and discriminant Δ. In fact, in an analogous way to what we have seen above, we can prove that it suffices to write down the inequalities concerning $\alpha \in \mathfrak{E}$ of rank $r_\alpha \leq r/2$ which bound μ. If we only look for non-rigid semi-stable sheaves, that is, with discriminant $\Delta \geq 1/2$, then these conditions are equivalent to the conditions (S) and (S*). If we introduce the function $\delta : \mathbb{Q} \to \mathbb{Q}$ given by

$$\delta(\mu) = \sup_{\substack{\alpha \in \mathfrak{E} \\ |\mu - \alpha| < 3}} P(-|\alpha - \mu|) - \Delta_\alpha$$

then the necessary and sufficient condition for the existence of non-exceptional stable sheaves can be also be written as

$$\Delta \geq \delta(\mu).$$

The function $\mu \mapsto \delta(\mu)$ is periodic with period 1. Consider, for $\alpha \in \mathfrak{E}$, the interval I_α of $\mathbb{Q}$ consisting of rational numbers μ such that $P(-|\alpha - \mu|) - \Delta_\alpha > 1/2$. An elementary arithmetic computation (cf. [13]) shows that the intervals I_α are disjoint and cover $\mathbb{Q}$. On the interval I_α, the function δ is also defined by

$$\delta(\mu) = P(-|\alpha - \mu|) - \Delta_\alpha.$$

The curve $\Delta = \delta(\mu)$ has the following appearance:

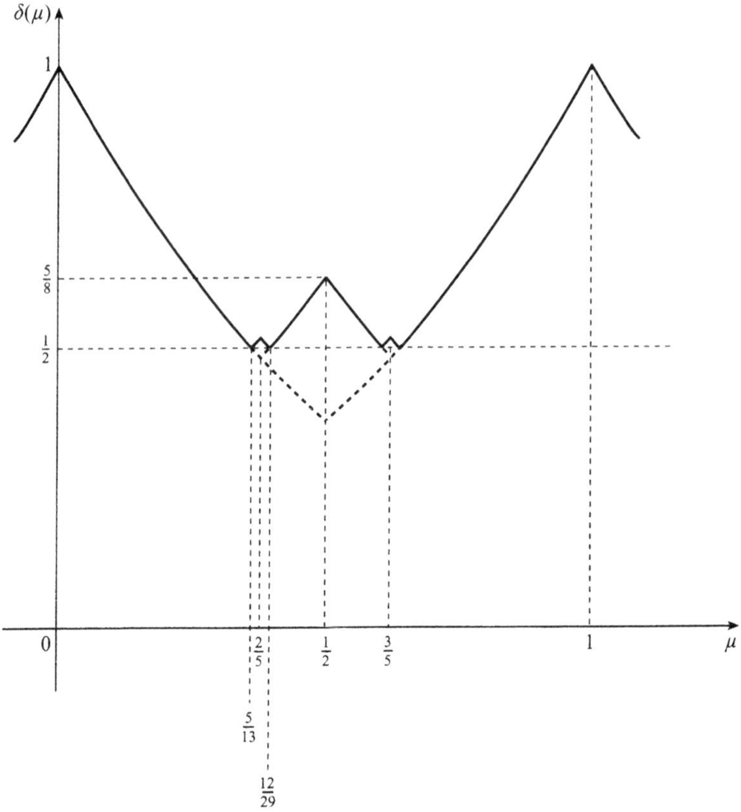

In particular, it follows that $\delta(\mu) > 1/2$.

Corollary 16.4.1 *There are no semi-stable sheaves on* $\mathbf{P}_2$ *with discriminant* $1/2$.

17

The Irreducibility

We aim to prove the following theorem [15].

Theorem 17.0.1 *The moduli space* $\mathbf{M} = \mathbf{M}_{\mathbf{P}_2}(r, c_1, c_2)$ *of semi-stable sheaves of rank* r *and Chern classes* c_1 *and* c_2 *on the projective plane is irreducible.*

We will assume that $r > 1$. The case $r = 1$ can be reduced to the study of the Hilbert scheme of points of $\mathbf{P}_2$ and must be treated separately, even if we can envisage a related method in order to treat it. The proof begins by showing that the open subset of locally-free polystable sheaves is dense: this proof is different to the one we have given in [15]; it consists in studying determinantal varieties, which allows us to describe the sheaves which are singular at a point. To show that this set is irreducible we project to a line from a centre a, after blowing-up the point, in order to reduce the problem to studying vector bundles on $\mathbf{P}_1$. The proof we give here is essentially the proof given by Ellingsrud [17] for stable bundles in the case $c_1 \not\equiv 0 \bmod r$. A simple modification allows us to treat the case $c_1 \equiv 0 \bmod r$. In the case $r = 2$, some direct proofs have been given by Barth [6] for c_1 even and by Hulek [32] for c_1 odd, by reducing the problem to the irreducibility of varieties of matrices. The case of semi-stable sheaves of rank 2 is also treated by Maruyama in [43]. The statement above has been extended recently by Hirschowitz and Laszlo [24] who show that the field of prioritary sheaves of fixed rank and degree is still irreducible.

Due to the fact that there is still no universal sheaf parametrized by $\mathbf{M}$, we need to work with a family parametrized by a suitable open subset of the Hilbert scheme. Let H be the Hilbert polynomial of sheaves of rank r and Chern classes (c_1, c_2), and let m be a large enough integer so that the conditions of section 14.4 are satisfied. In particular, for all semi-stable sheaves F with Hilbert polynomial H,

(i) the sheaf $F(m)$ is generated by its sections,
(ii) we have $H^q(F(m)) = 0$ for $q = 1$ and 2.

Consider the vector bundle $B = \mathbb{C}^{H(m)} \otimes \mathcal{O}_{\mathbf{P}_2}(-m)$. By construction, the moduli space $\mathbf{M} = \mathbf{M}_{\mathbf{P}_2}(r, c_1, c_2)$ of equivalence classes of semi-stable sheaves is the Mumford quotient of a smooth and equidimensional open subset $S = \mathbf{Hilb}^{ss}(B, H)$, of the Hilbert scheme $\mathbf{Hilb}(B, H)$; the Hilbert scheme of coherent sheaves which are quotients of B and have Hilbert polynomial H. This open subset S parametrizes a complete flat family F_s of semi-stable sheaves which are quotients of B. We aim to show that this variety is irreducible. It follows from this that the moduli space $\mathbf{M}$ is itself irreducible.

17.1 The density of the classes of locally-free sheaves

We start by examining the sheaves which are singular at a fixed point $a \in \mathbf{P}_2$.

Proposition 17.1.1 *Let $a \in \mathbf{P}_2$. The set of points $s \in S$ such that F_s is singular at a has codimension $r + 1$ if it is non-empty.*

For this we need to state some results about determinantal varieties.

Determinantal varieties

For a smooth algebraic variety Z consider a map of vector bundles of ranks m and n, respectively, over Z:

$$f : \mathcal{V} \longrightarrow \mathcal{W}.$$

Consider the sub-scheme Z^k of Z defined locally by the ideal of minors of order $m - k + 1$ whose underlying closed subset corresponds to the points $x \in X$ such that $\dim \ker f(x) \geq k$. This sequence of sub-schemes is decreasing. At a point $x \in Z^k \setminus Z^{k+1}$, the Zariski tangent space is the kernel of the natural Petri differential

$$d_x f : T_x X \longrightarrow L\big(\ker f(x), \operatorname{coker} f(x)\big).$$

This differential can be calculated in the following way. Locally trivialize the bundles $\mathcal{V}$ and $\mathcal{W}$ in a neighbourhood of x and compute the differential of the expression of f in the chosen trivialization. Then compose with the inclusion $\ker f(x) \hookrightarrow \mathcal{V}_x$ and the projection $\mathcal{W}_x \to \operatorname{coker} f(x)$. The linear map we then obtain does not depend on the choice of trivialization.

Lemma 17.1.2 *If the Petri differential surjects at the point x then the sub-scheme Z^k is smooth and has codimension $k(m-n+k)$ in a neighbourhood of x.*

Proof of Proposition 17.1.1 Let A be the kernel of the projection $\mathcal{O}_S \boxtimes B \to F$. This is a locally-free sheaf on $S \times X$ and, by flatness, we have an exact sequence

$$0 \longrightarrow A_s \xrightarrow{j_s} B \longrightarrow F_s \longrightarrow 0$$

for each point $s \in S$. Consider the induced map over $S \times \{a\}$

$$s \longmapsto j_s(a) : A|_{S\times\{a\}} \longrightarrow \mathcal{O}_S \otimes B(a).$$

Over s, the cokernel of $j_s(a)$ is isomorphic to $\operatorname{Tor}_1^\Lambda(F_s, \mathbb{C})$, where Λ is the local ring of $\mathbf{P}_2$ at the point a and so the injectivity of $j_s(a)$ means that the sheaf F_s is non-singular at the point a.

Lemma 17.1.3 *The Petri differential of j at $s \in S$ can be identified with the natural linear map*

$$T_sS \simeq \operatorname{Hom}(A_s, F_s) \longrightarrow L\big(\ker j_s(a), F_s(a)\big).$$

Consequently, if we can prove that this differential surjects then we could deduce that the associated strata are smooth and have the right dimension. In particular, the closed set of points s such that F_s is singular at a has codimension $r+1$, at least if it is not empty. To obtain the surjectivity it clearly suffices to prove the following lemma.

Lemma 17.1.4 *The sheaf $\mathcal{H}om(A_s, F_s)$ is generated by its global sections.*

Proof To prove this statement we consider the exact sequence of sheaves on X:

$$\mathcal{H}om(B, F_s) \longrightarrow \mathcal{H}om(A_s, F_s) \longrightarrow \mathcal{E}xt^1(F_s, F_s) \longrightarrow 0.$$

The choice of m guarantees that the sheaf $\mathcal{H}om(B, F_s)$ is generated by its global sections. It therefore suffices to check that the sheaf $\mathcal{E}xt^1(F_s, F_s)$ is itself generated by its global sections and its global sections come from $\mathrm{Hom}(A_s, F_s)$. The fact that $\mathcal{E}xt^1(F_s, F_s)$ is generated by its global sections is obvious because this sheaf has dimension 0 (in other words, it is supported at a finite number of points). From the relations between global Ext^1 and local $\mathcal{E}xt^1$ it follows that we have an exact sequence

$$\mathrm{Ext}^1(F_s, F_s) \longrightarrow H^0(\mathcal{E}xt^1(F_s, F_s)) \longrightarrow H^2(\mathcal{H}om(F_s, F_s)).$$

We know, from the choice of B, that $\mathrm{Ext}^1(B, F_s) = 0$. Therefore, the boundary map

$$\mathrm{Hom}(A_s, F_s) \longrightarrow \mathrm{Ext}^1(F_s, F_s)$$

surjects. The commuting diagram

$$\begin{array}{ccc} \mathrm{Hom}(A_s, F_s) & \longrightarrow & \mathrm{Ext}^1(F_s, F_s) \\ \downarrow & & \downarrow \\ H^0(\mathcal{H}om(A_s, F_s)) & \longrightarrow & H^0(\mathcal{E}xt^1(F_s, F_s)) \end{array}$$

shows that we are reduced to proving that $H^2(\mathcal{H}om(F_s, F_s)) = 0$. Put $E = F_s$ and consider the bidual $G = E^{**}$ of E. This is a locally-free sheaf which is still μ-semi-stable. The canonical map

$$\mathcal{H}om(E, E) \longrightarrow \mathcal{H}om(G, G)$$

injects and surjects away from a finite set of points and it induces an isomorphism at the level of H^2. So it suffices to check that

$$H^2(\mathcal{H}om(G, G)) = 0.$$

But this can be identified with $\mathrm{Ext}^2(G, G)$ and, since G is μ-semi-stable, it is zero by Serre duality. □

Corollary 17.1.5 *The set of points $s \in S$ such that F_s is locally-free is an open set whose complement, when it is non-empty, has codimension $r - 1$.*

Proof Consider the closed set Σ of pairs $(s, a) \in S \times \mathbf{P}_2$ such that the universal sheaf is singular at (s, a). Suppose that it is non-empty. Consider the projections

$$\begin{array}{ccc} \Sigma & \xrightarrow{\mathrm{pr}_2} & \mathbf{P}_2 \\ {\scriptstyle \mathrm{pr}_1}\downarrow & & \\ S & & \end{array}$$

Over $s \in S$, the fibre Σ_s represents the singular points of F_s and so the map $\Sigma \to S$ is finite. Over $\mathbf{P}_2$ the fibres are non-empty, pairwise isomorphic and of codimension $r+1$ in S. Hence, the image of Σ in S has codimension $r-1$. $\square$

Corollary 17.1.6 *If $r > 1$ then, in the moduli space $\mathbf{M} = \mathbf{M}_{\mathbf{P}_2}(r, c_1, c_2)$, the open set of points which represent locally-free polystable sheaves is dense.*

Proof of Lemma 17.1.3 The identification of the tangent space at s to the Hilbert scheme can be obtained in the following way. Over a neighbourhood W of (s, x) choose an isomorphism $\varphi : \mathrm{pr}_2^*(A_s) \simeq A$ which restricts to the identity on the fibre W_s over $s \in S$. The partial derivative $\partial_1(j\varphi)$ induces a map $T_sS \to H^0(W_s, \mathcal{H}om(A_s, F_s))$ which is independent of φ and the sheaf maps we get patch together when we change the open set W. Hence, we obtain an isomorphism $T_sS \simeq \mathrm{Hom}(A_s, F_s)$. By definition, the Petri differential of $j(a)$ restricted to the point a is induced by the partial differential $\partial_1(j\varphi)(s,a)$ and this clearly implies the stated result. $\square$

17.2 Restriction to a pencil of lines

Suppose that $r > 1$ and consider the open subset S' of $S = \mathbf{Hilb}^{ss}(B, H)$ corresponding to locally-free sheaves. This family of sheaves satisfies condition (P):

$$\mathrm{Ext}^2(F_s, F_s(-1)) = 0.$$

If d is a line then the family $(F_s|_d)_{s \in S}$ is complete which allows us to apply Corollary 15.4.3. This gives, depending on whether $c_1 \equiv 0 \bmod r$ or not,

Proposition 17.2.1 *Let d be a line of $\mathbf{P}_2$. Then the closed subset of points $s \in S$ such that the bundle $F_s|_d$ is not rigid has codimension 2 if $c_1 \not\equiv 0 \bmod r$ and has codimension 1 if $c_1 \equiv 0 \bmod r$.*

We now fix a pencil $\mathbf{P}_1 \subset |\mathcal{O}_{\mathbf{P}_2}(1)|$ of lines which pass through a given point of $\mathbf{P}_2$. Then we have the following consequence.

Corollary 17.2.2 *(i) If $c_1 \not\equiv 0 \bmod r$ then the open subset U of S' given by the points $s \in S'$ such that $F_s|_d$ is rigid for all $d \in \mathbf{P}_1$ is dense in S.*

(ii) Suppose that $c_1 \equiv 0 \bmod r$. The open subset U of S' defined by the points $s \in S'$ such that $F_s|_d$ is quasi-rigid for all lines $d \in \mathbf{P}_1$ and rigid for at least one line $d \in \mathbf{P}_1$, is dense in S.

Let U be the open subset defined in this corollary. We shall show that the fibres F_s of the family for $s \in U$ have a simple description.

Let $D \subset \mathbf{P}_1 \times \mathbf{P}_2$ denote the surface we get by blowing-up the point a. This is the "universal line" over $\mathbf{P}_1$. Consider the diagram

$$\begin{array}{ccc} D & \xrightarrow{p_2} & \mathbf{P}_2 \\ {\scriptstyle p_1}\downarrow & & \\ \mathbf{P}_1 & & \end{array}$$

The direct image $V = p_{1*}(p_2^*(\mathcal{O}(1))$ on $\mathbf{P}_1$ is a rank 2 vector bundle which is isomorphic to $\mathcal{O} \oplus \mathcal{O}(1)$ and the projective line bundle $\mathbf{P}(V)$ is isomorphic to D. On this projective line bundle the tautological invertible line bundle $\mathcal{O}_{\mathbf{P}(V)}(1)$ can be identified with the inverse image of $\mathcal{O}_{\mathbf{P}_2}(1)$ under the projection p_2. The inclusion $D \hookrightarrow \mathbf{P}_1 \times \mathbf{P}_2$ induces an isomorphism on their Picard groups and consequently an isomorphism

$$\mathbb{Z}^2 \simeq \mathrm{Pic}(D).$$

Let $\mathcal{O}(i,j)$ denote the invertible bundle associated to the pair $(i,j) \in \mathbb{Z}^2$. So the exceptional divisor e appears as the zero scheme of the canonical section of the invertible bundle $\mathcal{O}(-1,1)$.

Lemma 17.2.3 *Let $Y = D \times_{\mathbf{P}_1} D$. Consider the diagonal $\mathcal{D} \subset Y$. Then we have a resolution on Y*

$$0 \longrightarrow \mathcal{O}_D(-e) \boxtimes \mathcal{O}_D(0,-1) \longrightarrow \mathcal{O}_Y \longrightarrow \mathcal{O}_{\mathcal{D}} \longrightarrow 0.$$

Proof Let V_D be the pullback of V under the projection $p_1 : D \to \mathbf{P}_1$. Consider the canonical projection $V_D \to \mathcal{O}_D(0,1)$. This map gives rise to a map

$$V_D^* \boxtimes V_D \longrightarrow \mathcal{O}_D(-1,1) \boxtimes \mathcal{O}_D(0,1)$$

over Y. The identity of V corresponds to a canonical section of $\mathcal{O}_D(e) \boxtimes \mathcal{O}_D(0,1)$ whose zero scheme is exactly the diagonal. □

From now on we assume that $-r < c_1 \leq 0$.

We can always reduce to this case by twisting by a suitable invertible bundle. Consider the cartesian diagram

$$\begin{array}{ccc} Y & \xrightarrow{p_2} & D \\ {\scriptstyle p_1}\downarrow & & \downarrow \\ D & \longrightarrow & \mathbf{P}_1 \end{array}$$

For $s \in U$, we set $E = F_s$ and let E_D denote the pullback of E by the projection $p_2 : D \to \mathbf{P}_2$. Then we have

$$p_{1*}(E(-1)_D) = 0 \qquad \text{and} \qquad R^1 p_{1*}(E_D) = 0.$$

By applying the functor $- \otimes p_2^*(E_D)$ and then the functor p_{1*} to the exact sequence of the lemma above we get, after using the Base Change Theorem, a presentation of E_D as an extension:

$$0 \to p_1^*(p_{1*}(E_D)) \to E_D \to R^1 p_{1*}(E(-1)_D) \otimes_{\mathbf{P}_1} \mathcal{O}_D(-e) \to 0.$$

By the Semi-continuity Theorem, the sheaf $K = p_{1*}(E_D)$ is locally-free. This also follows for the sheaf $L = R^1 p_{1*}(E_D(-1))$ in the case $c_1 \neq 0$. If $c_1 = 0$ then the latter sheaf has dimension 0. The rank and degree of these sheaves can be easily computed using the Euler-Poincaré characteristic: if we set

$$n = -H(-1) = c_2 - \frac{1}{2} c_1(c_1 + 1)$$

then we get

	K	L
rank	$c_1 + r$	$-c_1$
degree	$-n$	$c_1 + n$

As s varies we obtain a family of locally-free sheaves K_s and a family of sheaves L_s which are also locally-free except if $c_1 = 0$ in which case the sheaves L_s have rank 0. □

Proposition 17.2.4 *The two families of sheaves K_s and L_s above are complete.*

Proof Consider a point $s \in U$ and the exact sequence on D

$$0 \longrightarrow K_s \otimes_{\mathbf{P}_1} \mathcal{O}_D \longrightarrow E_D \longrightarrow L_s \otimes_{\mathbf{P}_1} \mathcal{O}_D(-e) \longrightarrow 0.$$

We clearly have $\mathrm{Ext}^q(K_s \otimes_{\mathbf{P}_1} \mathcal{O}_D, L_s \otimes_{\mathbf{P}_1} \mathcal{O}_D(-e)) = 0$ for all q so that, in the notation of chapter 14, the vector space $\mathrm{Ext}^1_-(E_D, E_D)$ can be identified with $\mathrm{Ext}^1(E_D, E_D)$. We obtain a diagram

$$\begin{array}{ccc} T_sS & \xrightarrow{\omega} & \mathrm{Ext}^1(E_D, E_D) \\ & & \downarrow \\ & & \mathrm{Ext}^1(K_s, K_s) \oplus \mathrm{Ext}^1(L_s, L_s) \end{array}$$

in which the linear map ω denotes the Kodaira-Spencer infinitesimal deformation map. Moreover, we have

$$\mathrm{Ext}^2(L_s \otimes_{\mathbf{P}_1} \mathcal{O}(-e), K_s \otimes_{\mathbf{P}_1} \mathcal{O}_D) = 0,$$

which implies that the vertical map surjects. Our initial family of bundles F_s is complete and this is preserved under pullback by $p_2 : D \to \mathbf{P}_2$. Hence, the family of bundles E_D we obtain over D is still complete. In other words, the infinitesimal deformation map ω surjects. The composite maps $\omega_1 : T_s \to \mathrm{Ext}^1(K_s, K_s)$ and $\omega_2 : T_sS \to \mathrm{Ext}^1(L_s, L_s)$ are naturally Kodaira-Spencer infinitesimal deformation maps. These infinitesimal deformation maps are *a fortiori* surjective.

The end of the proof of Theorem 17.0.1

The case $c_1 \neq 0$

By Corollary 15.4.3, there exists an everywhere dense open subset $U' \subset U$ such that the vector bundles K_s and L_s are rigid for $s \in U'$. Since we know their rank and degree, these bundles are well defined up to isomorphism.

Fix two such rigid bundles which we still denote by K and L. In the vector space $\mathrm{Ext}^1(L, K \otimes_{\mathbf{P}_1} \mathcal{O}_D(e))$ which parametrizes extensions

$$0 \longrightarrow K \otimes_{\mathbf{P}_1} \mathcal{O}_D \longrightarrow G_w \longrightarrow L \otimes_{\mathbf{P}_1} \mathcal{O}_D(-e) \longrightarrow 0,$$

consider the open subset Ω given by the points $w \in \mathrm{Ext}^1(L, K \otimes_{\mathbf{P}_1} \mathcal{O}_D(e))$ which give rise to bundles G_w satisfying the following conditions:

(i) the bundle G_w is trivial on the exceptional divisor;
(ii) this bundle is the pullback by p_2 of a semi-stable bundle F_w on $\mathbf{P}_2$.

These bundles F_w clearly have rank r and Chern classes c_1 and c_2. Let m be the integer defined in the introduction of this chapter and consider the variety of frames R given by pairs (w, φ) consisting of a point $w \in \Omega$

and a trivialization $\varphi : \mathbb{C}^{H(m)} \simeq H^0(F_w(m))$. The universal property of the Hilbert scheme gives a map

$$f : R \longrightarrow S$$

whose image contains the open set U' we have already described above. The open set Ω is clearly irreducible and so R is irreducible. Consequently, U' is itself irreducible and so $\mathbf{M}$ is irreducible.

The case $c_1 = 0$

As above, we can find an open everywhere dense subset $U' \subset U$ such that K_s is rigid and hence well defined up to isomorphism. The sheaves L_s have rank 0 and degree n. Such a sheaf is isomorphic to the cokernel of a generically injective map

$$z : \mathbb{C}^n \otimes \mathcal{O}_{\mathbf{P}_1}(-1) \longrightarrow \mathbb{C}^n \otimes \mathcal{O}_{\mathbf{P}_1}.$$

Conversely, fix such a rigid sheaf K and consider the open set Z of maps z above which are generically injective. By considering the universal map on Z we get a cokernel $\mathcal{L}$ over $Z \times \mathbf{P}_1$ which is flat over Z and which induces the sheaf L_z over each point z. Consider the first projection $p_1 : Z \times D \to Z$ and the vector bundle

$$\mathcal{W} = \mathcal{E}xt^1_{p_1}(\mathcal{L} \otimes \mathcal{O}_D(-e), K \otimes_{\mathbf{P}_1} \mathcal{O}_D)$$

on Z, whose fibre over $z \in Z$ is the vector space which parametrizes extensions on D of the form

$$0 \longrightarrow K \otimes_{\mathbf{P}_1} \mathcal{O}_D \longrightarrow G_w \longrightarrow L_z \otimes_{\mathbf{P}_1} \mathcal{O}_D(-e) \longrightarrow 0.$$

Then we can consider, as in the case $c_1 \neq 0$, the open set $\Omega \subset \mathcal{W}$ of points $w \in \mathcal{W}$ such that the sheaf G_w is locally-free and satisfies conditions (i) and (ii). The varieties Z and Ω are irreducible so we can complete the proof exactly in the same way as for the case $c_1 \neq 0$. □

Remark.
This presentation is also very useful for rationality questions.

Theorem 17.2.5 (Hulek, Ellingsrud and Strømme, Maeda) *The moduli space* $\mathbf{M}_{\mathbf{P}_2}(2, c_1, c_2)$ *is a rational variety.*

It is essentially the presentation above which allows us to establish the rationality in the case where $4c_2 - c_1^2 \not\equiv 0 \bmod 8$ (Ellingsrud and Strømme

[18]). In particular, it allows us to treat the case where c_1 is odd, also studied by Hulek [32] using other methods. The proof is very much more delicate and we need to appeal to other descriptions if $4c_2 - c_1^2 \equiv 0 \bmod 8$: this is Maeda's result [42].

18

The Picard Group

Consider the moduli space $\mathbf{M} = \mathbf{M}_{\mathbf{P}_2}(r, c_1, c_2)$ of S-equivalence classes of semi-stable sheaves of rank r and Chern classes c_1 and c_2 on the projective plane. Let μ denote the slope and Δ the discriminant of these sheaves. If P is the Hilbert polynomial of the trivial rank 1 bundle then, as in section 16.4, we set

$$\delta(\mu) = \sup_{\substack{\alpha \in \mathfrak{E} \\ |\mu - \alpha| < 3}} P(-|\mu - \alpha|) - \Delta_\alpha.$$

The moduli space $\mathbf{M}$ is non-empty and not reduced to a point if and only if $\Delta \geq \delta(\mu)$. In which case, the moduli space $\mathbf{M}$ is an irreducible projective variety of dimension $r^2(2\Delta - 1) + 1$ and is a Mumford quotient of a smooth variety by the action of a reductive group, so it is normal. We aim to prove the following theorem [14].

Theorem 18.0.1 (Drézet) *(i) If $\Delta > \delta(\mu)$ then the Picard group of* $\mathbf{M}$ *is a free abelian group on two generators.*

(ii) If $\Delta = \delta(\mu)$ then the Picard group is an infinite cyclic group.

The proof which we shall give here is different from the initial proof of Drézet; it was given at the annual conference of EUROPROJ [39]. It rests on a weak version of the Brill-Noether Theorem proved for us by Göttsche and Hirschowitz [24]. It essentially allows us to show that the family of semi-stable sheaves for which we can establish existence forms a large open subset of $\mathbf{M}$. The proof can be adapted without difficulty to semi-stable torsion sheaves on the projective plane (cf. [39]).

18.1 Weak Brill-Noether

We know by the Serre Duality Theorem that if $\mu > -3$ then we have $H^2(F) = 0$ for all semi-stable sheaves F of rank r and Chern classes c_1 and c_2.

Theorem 18.1.1 (Göttsche and Hirschowitz [24]) *Suppose that $\mu > -3$ and let χ denote the Euler-Poincaré characteristic.*

(i) If $\chi > 0$ (respectively < 0) then the closed subset of $\mathbf{M}$ given by the classes of semi-stable sheaves F satisfying the condition $H^1(F) \neq 0$ (respectively $H^0(F) \neq 0$) has codimension at least 2.

(ii) If $\chi = 0$ then the sub-scheme of $\mathbf{M}$ given by the determinantal condition $H^1(F) \neq 0$ is an integral hypersurface, if it is non-empty.

This hypersurface is actually a Cartier divisor. More generally, a theorem of Drézet shows that the moduli space $\mathbf{M}$ is a locally factorial variety which means that all Weil divisors are Cartier divisors [14].

18.2 The homomorphism $\lambda_{\mathbf{M}} : K(\mathbf{P}_2) \to \mathrm{Pic}(\mathbf{M})$

Suppose that Y is a smooth algebraic variety. Then we let $K(Y)$ denote the Grothendieck algebra of classes of algebraic vector bundles on Y. In the following we will consider, in particular, the Grothendieck group $K(\mathbf{P}_2)$: this is a free abelian group of rank 3. For example, we obtain an isomorphism

$$K(\mathbf{P}_2) \simeq \mathbb{Z}^3$$

given by the rank r, the Chern class c_1 and the Euler-Poincaré characteristic χ. We have an integral quadratic form

$$u \longmapsto \chi(u^2) = 2r\chi + c_1^2 - r^2$$

on $K(\mathbf{P}_2)$. We let $\langle\,,\rangle$ denote the associated polar form. It is easily seen that this quadratic form has signature $(2,1)$. We also have an involution $u \mapsto u^*$ on $K(\mathbf{P}_2)$ which takes a class of vector bundles to its dual. This involution leaves the quadratic form $u \mapsto \langle u^*, u\rangle$ invariant. The dimension of the moduli space can be written as

$$\dim \mathbf{M}(r, c_1, c_2) = 1 - \langle c^*, c\rangle,$$

where $c \in K(\mathbf{P}_2)$ is the class of semi-stable sheaves of rank r and Chern classes c_1 and c_2. So the exceptional bundles are the stable bundles whose class $c \in K(\mathbf{P}_2)$ satisfies the condition $\langle c^*, c\rangle = 1$.

Consider an algebraic variety S equipped with an action of an algebraic group G and the diagram

$$\begin{array}{ccc} S \times \mathbf{P}_2 & \xrightarrow{p_2} & \mathbf{P}_2 \\ {\scriptstyle p_1}\downarrow & & \\ S & & \end{array}$$

The group G acts naturally on $S \times \mathbf{P}_2$. Let F be a coherent algebraic G-sheaf on $S \times \mathbf{P}_2$, that is, a sheaf with a G-action and which is S-flat. Consider, for $u \in K(\mathbf{P}_2)$, the invertible G-sheaf

$$\lambda_F(u) = \det p_{1!}(F(u))$$

on S. In this formula, $F(u)$ denotes the class in the Grothendieck group $K^G(S \times \mathbf{P}_2)$ of classes of coherent algebraic G-sheaves given by the pullback of u under p_2. The map

$$p_{1!} : K^G(S \times \mathbf{P}_2) \longrightarrow K^G(S)$$

is the map which takes the class of a coherent algebraic G-sheaf $\mathcal{F}$ to the class of $\sum_q (-1)^q R^q p_{1*}(\mathcal{F})$. So we obtain a group homomorphism

$$K(\mathbf{P}_2) \longrightarrow \mathrm{Pic}^G(S)$$

into the Picard group of invertible bundles on S which have a G action.

Lemma 18.2.1 *(i) The construction of $\lambda_F(u)$ is compatible with base change $S' \to S$.*

(ii) Given an exact sequence

$$0 \longrightarrow F' \longrightarrow F \longrightarrow F'' \longrightarrow 0$$

of coherent G-sheaves on $S \times \mathbf{P}_2$ flat over S, we have

$$\lambda_F(u) = \lambda_{F'}(u) \otimes \lambda_{F''}(u)$$

in $\mathrm{Pic}^G(S)$.

(iii) If A is an invertible G-bundle on S then we have

$$\lambda_{F \otimes p_1^*(A)}(u) = \lambda_F(u) \otimes A^{\otimes \langle c, u \rangle}.$$

We apply this to the case where S is the open subset $\mathbf{Hilb}^{ss}(B, H)$ of the Hilbert scheme $\mathbf{Hilb}(B, H)$ defined in section 14.3, equipped with the action of the algebraic group $G = \mathrm{Aut}(B)$ which we use to construct $\mathbf{M} = \mathbf{M}_{\mathbf{P}_2}(r, c_1, c_2)$. This open set is equipped with a universal quotient sheaf F. For each $u \in K(\mathbf{P}_2)$ we obtain an invertible G-bundle $\lambda_F(u)$ on S, well defined up to isomorphism. Let $\pi : S \to \mathbf{M}$ denote the canonical projection.

Theorem 18.2.2 *If $\langle c, u\rangle = 0$ then the invertible bundle $\lambda_F(u)$ gives rise to a well defined element of* Pic($\mathbf{M}$), *denoted $\lambda_{\mathbf{M}}(u)$.*

This element is characterized by the following universal property: for each family F' of semi-stable sheaves of rank r and Chern classes c_1 and c_2 parametrized by a smooth variety S', we have

$$f_{F'}^*(\lambda_{\mathbf{M}}(u)) = \lambda_{F'}(u),$$

where $f_{F'} : S' \to \mathbf{M}$ denotes the moduli map associated to F'.

The proof rests on a geometric invariant theory lemma, due to Kempf, Drézet and Narasimhan [16].

Lemma 18.2.3 *Let G be an algebraic group acting on an algebraic variety X and let $\pi : X \to Y$ be a good quotient of X by the action of G. The pullback map*

$$\pi^* : \mathrm{Pic}(Y) \longrightarrow \mathrm{Pic}^G(X)$$

injects and its image is the group of classes of invertible G-bundles L satisfying the following property: for each point $x \in X$ with closed orbit, the stabilizer G_x of x acts trivially on the fibre $L(x)$ of L over x.

Proof of Theorem 18.2.2 The points with closed orbits correspond to sheaves F_s which are polystable, in other words, are direct sums of stable sheaves of the same reduced Hilbert polynomial. Such a sheaf can be written as

$$F_s = F_1^{k_1} \oplus \cdots \oplus F_\ell^{k_\ell},$$

where the F_i are stable sheaves with the same reduced Hilbert polynomial and are pairwise non-isomorphic. At such a point $s \in S$, the stabilizer Stab(s) of the point s can be identified with the product group $\mathrm{GL}(k_1) \times \cdots \times \mathrm{GL}(k_\ell)$. Taking the above lemma into account, we can compute the action of the stabilizer on the fibre over s, in other words, the vector space $\lambda_{F_s}(u)$. If we let $c(F_i)$ denote the class of F_i in $K(\mathbf{P}_2)$ then we can see that the action is given by

$$(g_1, \ldots, g_\ell) \longmapsto \prod_{i=1,\ldots,\ell} (\det g_i)^{\langle c(F_i), u\rangle}$$

for $(g_1, \ldots, g_\ell) \in \mathrm{Stab}(s)$. Now, in $K(\mathbf{P}_2) \otimes \mathbb{Q}$, the fact that F and F_i have the same reduced Hilbert polynomial means that

$$\frac{c}{r} = \frac{c(F_i)}{r_i},$$

where r_i denotes the rank of F_i. Hence, the hypothesis implies that the action of the stabilizer is trivial. So we have obtained a construction of $\lambda_{\mathbf{M}}(u)$. The universal property of this invertible bundle clearly follows from the construction of the moduli map $f_{F'}$. □

Remark.
The statement of 18.2.2 above does not extend as such to other algebraic surfaces. To construct the invertible bundle $\lambda_{\mathbf{M}}(u)$ we need to impose more restrictive conditions on u [36]. However, we can construct it over the open set of stable points. This phenomenon allows us to explain why certain moduli spaces, for example on a quadric, are normal but are not locally factorial. For the projective plane, these moduli spaces are locally factorial [14].

Let $c^\perp$ denote the orthogonal sub-group to c in $K(\mathbf{P}_2)$. We then obtain a group homomorphism $u \mapsto \lambda_{\mathbf{M}}(u)$:

$$\lambda_{\mathbf{M}} : c^\perp \longrightarrow \mathrm{Pic}(\mathbf{M}).$$

We are going to use this map to make Drézet's Theorem more precise.

Definition 18.2.4 *Let $c \in K(\mathbf{P}_2)$ have rank $r > 0$, slope μ and discriminant Δ such that $\Delta = \delta(\mu)$. The* exceptional bundle associated *to c is the unique exceptional bundle with slope $\alpha \in \mathfrak{C}$ whose class a in $K(\mathbf{P}_2)$ satisfies the following conditions:*

$$-3 < \mu + \alpha \leq 0 \qquad \textit{and} \qquad \langle a, c \rangle = 0.$$

The existence of such an exceptional bundle is guaranteed by the definition of δ after taking duals and twisting by $\mathcal{O}(3)$. The uniqueness follows from the description of δ given in chapter 16.

The following statement is a more precise version of Theorem 18.0.1.

Theorem 18.2.5 *Let c be the class in $K(\mathbf{P}_2)$ of the coherent algebraic sheaves of rank $r > 0$ and Chern classes c_1 and c_2. Let μ denote the slope and Δ the discriminant.*

(i) If $\Delta > \delta(\mu)$ then the homomorphism $\lambda_{\mathbf{M}}$ above is an isomorphism.

(ii) If $\Delta = \delta(\mu)$ then the homomorphism $\lambda_{\mathbf{M}}$ is an epimorphism whose kernel is the sub-group generated by the class a of the exceptional bundle associated to c.

Recall that $\langle a^*, a\rangle = 1$. It follows, in case (ii), that $c^\perp$ is a direct sum of the sub-group generated by a and the orthogonal complement of a^* in $c^\perp$. This orthogonal complement is an infinite cyclic group isomorphic to the quotient group $c^\perp/(a)$. This leads to 18.0.1.

18.3 The codimension of the Shatz strata

This section gives some additional information about the Shatz strata of a complete flat family of torsion-free sheaves on the projective plane whose discriminant satisfies the inequality

$$\Delta > \delta(\mu).$$

Lemma 18.3.1 *Let F_s be a complete family of torsion-free coherent sheaves on the projective plane with rank r, slope μ and discriminant Δ, parametrized by a smooth variety S. Suppose that $\Delta > \delta(\mu)$ and such that the following condition (P) is satisfied for each $s \in S$:*

$$\text{(P)} \qquad \operatorname{Ext}^2(F_s, F_s(-1)) = 0.$$

Then the complement of the open set of points $s \in S$ such that F_s are stable has codimension at least 2 in S.

Proof We first show that the closed set of points $s \in S$ such that F_s are unstable has codimension at least 2. We need to show that the Shatz strata have codimension at least 2. By Corollary 15.4.4, we know that we have the condition

$$\text{(R)} \qquad \mu_{\max}(F_s) - \mu_{\min}(F_s) < 3$$

away from a closed set of codimension ≥ 2. By Corollary 15.4.5 it suffices to prove that if (R) is satisfied for F_s and if F_s has a Harder-Narasimhan filtration of length $\ell \geq 2$ then we have

$$\sum_{i<j} r_i r_j(\Delta_i + \Delta_j - P(\mu_j - \mu_i)) > 1,$$

where r_i, μ_i and Δ_i denote the rank, slope and discriminant, respectively, of the Harder-Narasimhan grading $\operatorname{gr}_i(F_s)$.

Recall the identity

$$\chi(\mathrm{gr}_i, \mathrm{gr}_j) = r_i r_j (P(\mu_j - \mu_i) - \Delta_i - \Delta_j)$$

so that condition (R) implies that each of the terms of the sum above is positive. Suppose that the sum equals 1. We distinguish two cases.

(a) If $\chi(\mathrm{gr}_1, \mathrm{gr}_\ell) = 0$ then the arguments we used in section 16.2 show that one of the sheaves gr_1 and gr_ℓ is semi-exceptional because $0 \leq \mu_1 - \mu_\ell < 3$. Then we have

$$\begin{aligned} \chi(\mathrm{gr}_1, F_s) &= \chi(\mathrm{gr}_1, \mathrm{gr}_1) - \sum_{i>1} \chi(\mathrm{gr}_1, \mathrm{gr}_i) \\ &\geq \chi(\mathrm{gr}_1, \mathrm{gr}_1) - 1 \geq 0 \end{aligned}$$

which contradicts the hypothesis that $\Delta > \delta(\mu)$. We treat the case where gr_ℓ is semi-exceptional in the same way, by studying $\chi(F_s, \mathrm{gr}_\ell)$.

(b) Suppose that $\chi(\mathrm{gr}_1, \mathrm{gr}_\ell) = -1$, that is,

$$\Delta_1 + \Delta_\ell = \frac{1}{r_1 r_\ell} + P(\mu_1 - \mu_\ell).$$

The computation that we are going to give shows that neither gr_1 nor gr_ℓ can be semi-exceptional. Since the discriminant of a semi-stable sheaf is not equal to 1/2 we have that

$$r_1^2(2\Delta_1 - 1) \geq 1 \qquad \text{and} \qquad r_\ell^2(2\Delta_\ell - 1) \geq 1$$

by section 16.4, and hence

$$\frac{1}{r_1 r_\ell} + P(\mu_\ell - \mu_1) \geq 1 + \frac{1}{2}\left(\frac{1}{r_1^2} + \frac{1}{r_\ell^2}\right).$$

This can be written as

$$P(\mu_\ell - \mu_1) - 1 \geq \frac{1}{2}\left(\frac{1}{r_1} - \frac{1}{r_\ell}\right)^2,$$

which implies, in view of $0 \leq \mu_1 - \mu_\ell < 3$, that $\mu_1 = \mu_\ell$ and $r_1 = r_\ell$. But the inequality cannot be strict, which implies

$$\Delta_1 = \Delta_\ell = \frac{1}{2}\left(1 + \frac{1}{r_1^2}\right).$$

But this contradicts the definition of the Harder-Narasimhan filtration. Hence, neither case (a) nor case (b) can occur.

The proof of Corollary 15.4.6 then shows that the complement of the open set of stable points has codimension greater than 1 except perhaps if there exists in the family of semi-stable sheaves F_s whose Jordan-Hölder grading $\mathrm{gr}(F_s) = \bigoplus_{i=1}^{\ell} \mathrm{gr}_i(F_s)$ satisfy the condition

$$(2\Delta - 1) \sum_{i<j} r_i r_j = 1.$$

But this implies that $\ell = 2$, $r_1 = r_2$ and $(2\Delta - 1)r_1^2 = 1$. Thus the moduli space of stable sheaves of rank r_1, slope μ and discriminant Δ would be non-empty and dimension 2. These moduli spaces have been classified by Drézet [13, proposition 34] we can only obtain them when $\Delta = \delta(\mu)$. This completes the proof. □

18.4 The large open set U

After twisting the sheaves by a suitable invertible bundle we can assume that the following conditions hold:

$$\mu > 0, \qquad H(-2) < 0 \qquad \text{and} \qquad H(-1) \geq 0.$$

These imply that $\chi = H(0) > 0$. Consider the vector spaces K_0, K_1 and K_2 of dimensions $H(0)$, $H(-1)$ and $-H(-2)$, respectively. Let $U \subset \mathbf{M}$ be the open set of equivalence classes of semi-stable sheaves whose Jordan-Hölder grading F satisfies the following conditions:

$$H^1(F) = 0, \qquad H^1(F(-1)) = 0 \qquad \text{and} \qquad H^0(F(-2)) = 0.$$

By the result of Göttsche and Hirschowitz, the complement of this open set has codimension 2 if $H(-1) \neq 0$. If $H(-1) = 0$ then the determinantal variety given by $H^1(F(-1)) \neq 0$ is an integral hypersurface, if it is non-empty. For such a sheaf F we then have

$$h^1(F(-2)) = -H(-2), \quad h^0(F(-1)) = H(-1) \quad \text{and} \quad h^0(F) = H(0).$$

It follows easily from the Beilinson spectral sequence that such a sheaf F is the cokernel of a map of vector bundles

$$\mathcal{V} = K_2 \otimes \mathcal{O}_{\mathbf{P}_2}(-1) \oplus K_1 \otimes Q^* \longrightarrow \mathcal{W} = K_0 \otimes \mathcal{O}_{\mathbf{P}_2}$$

which injects away from a finite number of points. This family of sheaves can also be used to establish the existence theorem of semi-stable sheaves.

Conversely, consider the family $\mathcal{M}$ of maps $s : \mathcal{V} \to \mathcal{W}$ which injects except perhaps at a finite number of points. The complement in the vector space $\mathrm{Hom}(\mathcal{V}, \mathcal{W})$ of this open set is a closed set of codimension $r + 1$. The cokernel of such a map s is then a sheaf of depth at least

1 at each point and whose singularity set has dimension 0. It follows from a well known criterion due to Serre that this sheaf $F_s = \operatorname{coker} s$ is torsion-free. As in section 16.2, this family of torsion-free sheaves satisfies condition (P) and so the complement of the open set $\mathcal{M}^{ss}$ of points s such that F_s is semi-stable and such that the Jordan-Hölder grading $\operatorname{gr}(F_s)$ is contained in U has codimension at least 2 if $\Delta > \delta(\mu)$ and has codimension at least 1 if $\Delta = \delta(\mu)$ by section 18.3. Then the coarse moduli property gives us a map

$$\pi : \mathcal{M}^{ss} \longrightarrow U$$

which is equivariant under the natural action of the algebraic group $G = \operatorname{Aut}(\mathcal{V}) \times \operatorname{Aut}(\mathcal{W})$.

Lemma 18.4.1 *The map π makes the open set U into a good quotient of $\mathcal{M}^{ss}$ under the action of G.*

The proof rests on the transitivity properties of good quotients.

Lemma 18.4.2 [56] *Let G be an algebraic group acting on an algebraic variety X and N a distinguished closed sub-group. Consider a good quotient $f : X \to Y$ of X by N and a G-equivariant map $g : X \to Z$ which factorizes as in the diagram*

$$\begin{array}{ccc} X & \xrightarrow{\ f\ } & Y \\ & {\scriptstyle g}\searrow & \downarrow {\scriptstyle h} \\ & & Z \end{array}$$

where h is (G/N)-equivariant. Then g is a good quotient if and only if h is a good quotient.

Proof of Lemma 18.4.1 We begin by remarking that the group $\operatorname{Aut}(\mathcal{V})$ contains the distinguished sub-group N given by the group of automorphisms of $\mathcal{V}$ which can be written as

$$\begin{pmatrix} 1 & 0 \\ g & 1 \end{pmatrix}$$

with $g \in \operatorname{Hom}(K_2 \otimes \mathcal{O}_{\mathbf{P}_2}(-1), K_1 \otimes Q^*)$. This group acts properly and freely on $\mathcal{M}^{ss}$. There exists a good quotient $\mathcal{M}^{ss} \to \mathcal{N}^{ss}$ which is a locally trivial principal bundle with structure group N. The variety $\mathcal{N}^{ss}$ consists of pairs (u, v), where $u : K_1 \otimes Q^* \to K_0 \otimes \mathcal{O}_{\mathbf{P}_2}$ is a generically injective map and $v : K_2 \otimes \mathcal{O}_{\mathbf{P}_2}(-1) \to \operatorname{coker} u$ is a sheaf map which is also generically injective such that the cokernel of v is semi-stable. Since the dimension of the vector space

$$\operatorname{Hom}(K_2 \otimes \mathcal{O}_{\mathbf{P}_2}(-1), \operatorname{coker} u)$$

remains constant as u varies, the variety $\mathcal{N}^{ss}$ is an open subset of a vector bundle over an open subset of $\operatorname{Hom}(K_1 \otimes Q*, K_0 \otimes \mathcal{O}_{\mathbf{P}_2})$. Choosing the integer m sufficiently large as in chapter 17, consider the inverse image V of U in the Hilbert scheme $\mathbf{Hilb}(B, H)$. Let F be the universal quotient sheaf on $V \times \mathbf{P}_2$. Consider the frame bundle R of quadruples $(s, (\kappa_2, \kappa_1, \kappa_0))$, where $s \in V$ and

$$\kappa_2 : K_2 \simeq H^1\big(F_s(-2)\big) \text{ and } \kappa_i : K_i \simeq H^0\big(F_s(-i)\big) \text{ if } i = 0 \text{ and } i = 1.$$

Consider the commuting diagram

$$\begin{array}{ccc} R & \xrightarrow{\sigma} & \mathcal{N}^{ss} \\ {\scriptstyle\rho}\downarrow & & \downarrow{\scriptstyle p_2} \\ V & \xrightarrow{p_1} & U \end{array}$$

in which the map σ is given by the obvious universal property of the variety $\mathcal{N}^{ss}$. The variety R is equipped with an action of

$$\operatorname{GL}(H(m)) \times \prod_i \operatorname{GL}(K_i).$$

The map ρ is a locally trivial principal bundle with structure group $\prod_i \operatorname{GL}(K_i)$. The map σ is also a locally trivial principal bundle with structure group $\operatorname{GL}(H(m))$ and so these are good quotients. We know that p_1 is a good quotient under the action of $\operatorname{GL}(H(m))$ and by the transitivity properties of good quotients stated in Lemma 18.4.2 above, it follows that p_2 is a good quotient under the action of $\prod_i \operatorname{GL}(K_i)$. Composing with the projection $\mathcal{M}^{ss} \to \mathcal{N}^{ss}$ we see, by applying Lemma 18.4.2 again, that π is a good quotient under the action of G. □

18.5 The end of the calculation

An exact sequence

Consider an algebraic group G acting on an algebraic variety X (assumed to be integral for simplicity). Let $\mathcal{O}^*(X)$ denote the multiplicative group of invertible regular functions. A *crossed morphism* $\varphi : G \to \mathcal{O}^*(X)$ is a map of varieties $\varphi : G \times X \to \mathbb{C}^*$ such that for $g, g' \in G$ and $x \in X$ we have $\varphi(gg', x) = \varphi(g, g'x)\varphi(g', x)$. Such a crossed morphism φ is said to be *principal* if there exists $\psi \in \mathcal{O}^*(X)$ such that, for $(g, x) \in G \times X$,

$$\varphi(g, x) = \frac{\psi(gx)}{\psi(x)}.$$

We have an exact sequence

$$0 \longrightarrow H^1(G, \mathcal{O}^*(X)) \longrightarrow \mathrm{Pic}^G(X) \longrightarrow \mathrm{Pic}(X)^G,$$

where $H^1(G, \mathcal{O}^*(X))$ is the quotient of the crossed morphisms $G \to \mathcal{O}^*(X)$ by the sub-group of the principal crossed morphisms and $\mathrm{Pic}(X)^G$ is the sub-group of the Picard group of elements which are invariant under the G action. If $\mathcal{O}^*(X) = \mathbb{C}^*$ then the group of crossed morphisms can be identified with the character group $G \to \mathbb{C}^*$.

The group $\mathrm{Pic}^{\overline{G}}(\mathcal{M}^{ss})$

We return to the action of the group $G = \mathrm{Aut}(\mathcal{V}) \times \mathrm{Aut}(\mathcal{W})$ on the variety $\mathcal{M}$. Consider the action of the quotient $\overline{G} = G/\mathbb{C}^*$ of G by the sub-group $\mathbb{C}^*$ which is contained in the centre of G.

Lemma 18.5.1 *Let F be the universal family on $\mathcal{M} \times \mathbf{P}_2$ parametrized by $\mathcal{M}$.*

(i) The map $\lambda_F : K(\mathbf{P}_2) \to \mathrm{Pic}^G(\mathcal{M})$ is an epimorphism and an isomorphism if $H(-1) \neq 0$.

(ii) This homomorphism induces an epimorphism

$$c^{\perp} \longrightarrow \mathrm{Pic}^{\overline{G}}(\mathcal{M}).$$

This epimorphism is an isomorphism if $H(-1) \neq 0$.

Proof We know that $\mathcal{M}$ is the complement of a closed set of codimension at least 2 in a vector space and so

$$\mathcal{O}^*(\mathcal{M}) = \mathbb{C}^* \qquad \text{and} \qquad \mathrm{Pic}(\mathcal{M}) = 0.$$

Consequently, the group $\mathrm{Pic}^G(\mathcal{M})$ is isomorphic to the character group of G. The projection $G \to \prod_{i=0,1,2} \mathrm{GL}(K_i)$ induces an isomorphism

$$\mathrm{Char}(G) \simeq \prod_{i=0,1,2} \mathrm{Char}(\mathrm{GL}(K_i))$$

and so this character group is $\mathbb{Z}^2$ or $\mathbb{Z}^3$ depending on whether $H(-1)$ is zero or not. When K_i is not zero then we can take the determinant map $\mathrm{GL}(K_i) \to \mathbb{C}^*$ to be the generator of $\mathrm{Char}(\mathrm{GL}(K_i))$. Consider the diagram

$$\begin{array}{ccc} \mathcal{M} \times \mathbf{P}_2 & \xrightarrow{p_2} & \mathbf{P}_2 \\ {\scriptstyle p_1}\downarrow & & \\ \mathcal{M} & & \end{array}$$

For $i = 0, 1, 2$, let e_i be the class in $K(\mathbf{P}_2)$ of the bundle $\mathcal{O}(-i)$. These classes form a basis of the Grothendieck group $K(\mathbf{P}_2)$. The element $p_{1*}(F(e_i))$ is the class in $K^G(\mathcal{M})$ of the vector-G bundle $\mathcal{M} \times K_i$ for $i = 0$ and $i = 1$. The class of $p_{1*}(F(e_2))$ is the negative of the class of the vector-G bundle $\mathcal{M} \times K_2$. It follows that the invertible G-bundle $\lambda_F(e_i)$ is the invertible bundle which comes from the generator of $\mathrm{Char}(\mathrm{GL}(K_i))$ for $i = 0$ and $i = 1$ and of the negative of this generator for $i = 2$ via the projection $G \to \mathrm{GL}(K_i)$. This implies assertion (i).

Then we have a commutative diagram

$$\begin{array}{ccc} K(\mathbf{P}_2) & \longrightarrow & \mathrm{Char}(G) \\ & {\scriptstyle (1)}\searrow & \downarrow{\scriptstyle (2)} \\ & & \mathbb{Z} \end{array}$$

in which the map (1) is given by $u \mapsto \langle c, u \rangle$ and the vertical map is induced by the inclusion $\mathbb{C}^* \hookrightarrow G$. This can be checked directly on the basis elements e_i. The kernel of (1) is, by definition, $c^\perp$ and the kernel of (2) is the group $\mathrm{Char}(\overline{G})$. The result now follows. □

Proof of Theorem 18.0.1

Consider the diagram

$$\begin{array}{ccc} \mathcal{M}^{ss} & \hookrightarrow & \mathcal{M} \\ {\scriptstyle \pi}\downarrow & & \\ U & \hookrightarrow & \mathbf{M} \end{array}$$

From the universal property of $\lambda_{\mathbf{M}}(u)$ it follows that this diagram induces a commutative diagram

$$\begin{array}{ccccc} c^\perp & & \xrightarrow{\lambda_F} & & \mathrm{Pic}^{\overline{G}}(\mathcal{M}) \\ {\scriptstyle \lambda_{\mathbf{M}}}\downarrow & & & & \downarrow{\scriptstyle (3)} \\ \mathrm{Pic}(\mathbf{M}) & \xrightarrow{\rho} & \mathrm{Pic}(U) & \xrightarrow{(4)} & \mathrm{Pic}^{\overline{G}}(\mathcal{M}^{ss}) \end{array}$$

We know from the preceding lemma that the map λ_F surjects and that the map (3) is obviously surjective. Moreover, the fact that U is a good quotient of $\mathcal{M}^{ss}$ implies that the map (4) injects. It follows that (4) is an isomorphism. To complete the proof we need to distinguish the cases $\Delta > \delta(\mu)$ and $\Delta = \delta(\mu)$.

The case $\Delta > \delta(\mu)$

By Lemma 18.3.1, the complement of $\mathcal{M}^{ss}$ has codimension at least 2 in $\mathcal{M}$. It follows that the map (3) is an isomorphism.

(a) If $H(-1) \neq 0$ then the map λ_F is an isomorphism. The same follows from ρ by Theorem 18.1.1. Hence, $\lambda_{\mathbf{M}}$ is an isomorphism.

(b) If $H(-1) = 0$ then the class e_1 of $\mathcal{O}(-1)$ is contained in $c^{\perp}$. The kernel of λ_F is a sub-group generated by e_1. The quotient $c^{\perp}/(e_1) \simeq \mathrm{Pic}(U)$ is isomorphic to $\mathbb{Z}$. Since e_1 is contained in $c^{\perp}$ we can consider the invertible bundle $\lambda_{\mathbf{M}}(-e_1)$. This invertible bundle is then equipped with a section whose zero set is non-empty by the following lemma and is exactly the integral hypersurface given by the classes of the sheaves E satisfying the determinantal condition $H^1(E(-1)) \neq 0$. This hypersurface coincides with the complement of U, away from a closed set of codimension at least 2.

Lemma 18.5.2 *Suppose $\mu > 0$, $H(-1) = 0$ and $\Delta > \delta(\mu)$. Then the determinantal sub-variety of $\mathbf{M}$ defined by the condition $H^1(E(-1)) \neq 0$ is non-empty.*

It follows that the element $\lambda_{\mathbf{M}}(-e_1)$ of $\mathrm{Pic}(\mathbf{M})$ which generates the kernel of the map ρ is not a torsion element. Hence, $\lambda_{\mathbf{M}}$ induces an isomorphism $\ker \lambda_F \to \ker \rho$. So $\lambda_{\mathbf{M}}$ is an isomorphism.

Proof Since $\Delta > \delta(\mu)$, we must have $\mu > 1$, so $c_1 > r$. For $r = 1$, $c_1 = 2$ we then have $c_2 = 3$ and the statement is obvious. It suffices to take $E = \mathcal{I}(2)$, where $\mathcal{I}$ is the ideal of three collinear points in the plane.

In view of this, we may assume that $c_1 \geq 3$. Consider the family $\mathcal{H}$ of maps

$$s : \mathcal{O}(-2) \oplus \mathcal{O}(-1)^{c_1-3} \longrightarrow \mathcal{O}^{c_1+r-3} \oplus \mathcal{O}(1)$$

which inject except perhaps at a finite number of points. The cokernel of such a map s gives rise to a torsion-free coherent sheaf E_s of rank r

and Hilbert polynomial H. This family $\mathcal{H}$ is not complete but the image of the Kodaira-Spencer infinitesimal deformation map

$$\omega : T_s\mathcal{H} \longrightarrow \operatorname{Ext}^1(E_s, E_s)$$

has codimension at most 1. Moreover condition (P) of Lemma 18.3.1 is satisfied. Then an obvious variation of Lemma 18.3.1 shows that the codimension of the Shatz strata is at least 1. Hence the open set $\mathcal{H}^{ss}$ of maps such that the cokernel E_s is semi-stable is everywhere dense.

The case $\Delta = \delta(\mu)$

We can go through the same argument but this time it only gives the surjectivity of $\lambda_{\mathbf{M}}$. Consider the exceptional bundle E_α associated to c whose slope is α and whose class in $K(\mathbf{P}_2)$ we denote by a. We can check immediately that if $\mathcal{F}$ is a family of semi-stable sheaves parametrized by a smooth algebraic variety S then we have, in the notation of section 18.2, $R^q p_{1*}(E_\alpha \otimes \mathcal{F}) = 0$ for all q from which we get $p_{1!}(E_\alpha \otimes \mathcal{F}) = 0$. This implies that $\lambda_{\mathbf{M}}(a) = 0$ which means that a is contained in the kernel of $\lambda_{\mathbf{M}}$.

So $\lambda_{\mathbf{M}}$ induces an epimorphism

$$c^{\perp}/(a) \longrightarrow \operatorname{Pic}(\mathbf{M}).$$

We have already seen that the left hand side is a cyclic group. Since $\mathbf{M}$ is a projective variety, its Picard group cannot be a finite group. Hence this homomorphism is an isomorphism. □

References

[1] Atiyah, M.F. *Vector bundles on an elliptic curve*, Proc. London Math. Soc. **7** (1959) 414–452.

[2] Atiyah, M.F. *K-Theory, Lecture notes by D.W. Anderson, Harvard 1964*, New York (1967).

[3] Atiyah, M.F. and Bott, R. *The Yang-Mills equations over Riemann surfaces*, Phil. Trans. Roy. Soc. London **A 308** (1982) 523–615.

[4] Atiyah, M.F. and MacDonald, I.G. *Introduction to commutative algebra*, Addison-Wesley (1969).

[5] Barth, W. *Some properties of rank-2 bundles on* $\mathbf{P}_n$, Math. Annalen **226** (1976) 125–150.

[6] Barth, W. *Moduli of vector bundles on the projective plane*, Invent. Math. **42** (1977) 63–91.

[7] Beilinson, A.A. *Coherent sheaves on* $\mathbf{P}_n$ *and problems in linear algebra*, Funkt. Analiz. Appl. **12 (3)** (1978) 68–69.

[8] Bogomolov, F.A. *Holomorphic tensors and vector bundles on projective varieties*, Math. USSR Izvestja **13** (1979) 499-555.

[9] Bourbaki, N. *Commutative algebra*, Chapter III.

[10] Donaldson, S.K. *A new proof of a theorem of Narasimhan and Seshadri*, J. Diff. Geom. **18** (1983) 269–277.

[11] Douady, R. and A., *Algèbre et théories galoisiennes*, (1977) Cedic, Fernand Nathan.

[12] Drézet, J.-M. *Fibrés exceptionnels et suite spectrale de Beilinson généralisée*, Math. Ann. **275** (1986) 25–48.

[13] Drézet, J.-M. *Fibrés exceptionnels et variétés de modules de faisceaux semi-stables sur* $\mathbf{P}_2$, J. reine angew. Math **380** (1987) 14–58.

[14] Drézet, J.-M. *Groupe de Picard des variétés de modules de faisceaux semi-stable sur* $\mathbf{P}_2(\mathbb{C})$, Ann. de l'Institut Fourier **38** (1988) 105–168.

[15] Drézet, J.-M. and Le Potier, J. *Fibrés stables et fibrés exceptionnels sur le plan projectif*, Ann. scient. Ec. Norm. Sup. **4e série, 18** (1985) 193–244.

[16] Drézet, J.-M. and Narasimhan, M.S. *Groupe de Picard des variétés de modules de fibrés semi-stables sur les courbes algébriques*, Invent. Math. **97** (1989) 53–94.

[17] Ellingsrud, G. *Sur l'irréductibilité du module des fibrés stables sur* $\mathbf{P}_2$, Math. Z. **182** (1983) 189–192.

[18] Ellingsrud, G. and Strømme, S.A. *On the rationality of the moduli space for rank 2 vector bundles on* $\mathbf{P}_2$, in *Singularities, representations of algebras and vector bundles*, Lecture Notes in Mathematics **1273**, Springer-Verlag (1985).

[19] Ellingsrud, G. and Strømme, S.A. *Towards the Chow ring of the Hilbert scheme of* $\mathbf{P}_2$, J. reine angew. Math. **441** (1993) 33–44.

[20] Flenner, H. *Restrictions of semistable bundles on projective varieties*, Comment. Math. Helv. **59** (1984) 635–650.

[21] Forster, O., Hirschowitz, A. and Schneider, M. *Type de scindage généralisé pour les fibrés stables*, in *Vector bundles and differential equations*, Birkhäuser (1980) 65–81.

[22] Gieseker, D. *On the moduli of vector bundles on an algebraic surface*, Ann. of Maths. **106** (1977) 45–60.

[23] Godement, R. *Topologie algébrique et théorie des faisceaux*, Paris (1958).

[24] Göttsche, L. and Hirschowitz, A., *Weak Brill-Noether for vector bundles on the projective plane*, Seminar given at the annual EUROPROJ conference (Catania, 1993), to appear.

[25] Gramain, A. *Topologie des surfaces*, Presses Universitaires de France, Paris (1971).

[26] Grauert, H. and Mülich, G. *Vektorbündel von Rang 2 über dem n-dimensionalen komplex projektiven Raum*, Manuscripta Math. **16** (1975) 75–100.

[27] Greenberg, M. *Lectures on algebraic geometry*, Mathematical Lecture Notes Series, Springer-Verlag (1967).

[28] Grothendieck, A. *Sur quelques points d'algèbre homologique*, Tôhoku Math. J. **9** (1957) 119–221.

[29] Grothendieck, A. *Techniques de construction et théorèmes d'existence en géométrie algébrique, IV: les schémas de Hilbert*, Sém. Bourbaki **221** (1960).

[30] Harder, G. and Narasimhan, M.S. *On the cohomology of moduli spaces of vector bundles on curves*, Math. Ann. **212** (1975) 215–248.

[31] Hartshorne, R. *Algebraic Geometry*, Graduate Texts in Mathematics, Springer-Verlag, New York (1977).

[32] Hulek, K. *Stable rank-2 vector bundles on* $\mathbf{P}_2$ *with* c_1 *odd*, Math. Ann. **242** (1979) 241–266.

[33] Husemoller, D. *Fibre bundles*, Springer-Verlag (1966).

[34] Kraft, H. *Geometrische Methoden in der Invariantentheorie*, Vieweg-Verlag, (1985).

[35] Lang, S. *Algebra*, Addison-Wesley (1984).

[36] Le Potier, J. *Fibré déterminant et courbes de saut sur les surfaces algébriques*, in *Complex projective geometry*, London Mathematical Society, Bergen (1989) 213–240.

[37] Le Potier, J. *Variété de modules de faisceaux semi-stable de rang élevé sur* $\mathbf{P}_2$, Proceedings of Symposia in Pure Mathematics Bowdoin, 1985 **46** (1987) 87–100.

[38] Le Potier, J. *L'espace de module de Simpson*, Exposé du 27 février 1992, Séminaire de géométrie algébrique, Jussieu.

[39] Le Potier, J. *Problème de Brill-Noether et groupe de Picard de l'espace de module des faisceaux semi-stables sur le plan projectif*, EUROPROJ 1993, Proceedings of the Annual Conference, Catania, to appear.

[40] Le Potier, J. *A propos de la construction de l'espace de modules des faisceaux semi-stable sur le plan projectif*, Bull. de la S.M.F. **122** (1994) 101–107

[41] Le Potier, J. *Faisceaux semi-stables et systèmes cohérents*, Proceedings of Durham Symposium 1993, to appear.

[42] Maeda, T. *An elementary proof of the rationality of the moduli space for rank 2 vector bundles on* $\mathbf{P}_2$, Hiroshima Math. J. **20** (1990) 103–107.

[43] Maruyama, M. *Moduli of stable sheaves, II*, J. Math. Kyoto University **18** (1978) 557–614.

[44] Maruyama, M. *The theorem of Grauert-Mülich-Spindler*, Math. Ann. **255** (1981) 317–533.

[45] Maruyama, M. *On the boundedness of families of torsion-free sheaves*, J. Math. Kyoto University **21-4** (1981) 673–701.

[46] Mehta, V.B. and Ramanathan, A. *Semistable sheaves on projective varieties and their restriction to curves*, Math. Ann. **258** (1982) 213–224.

[47] Mumford, D. and Fogarty, J. *Geometric invariant theory*, Springer-Verlag (1982).

[48] Narasimhan, M.S. and Ramanan, S. *Moduli of vector bundles on a compact Riemann surface*, Ann. of Maths. **89** (1969) 14–51.

[49] Narasimhan, M.S. and Seshadri, C.S. *Stable and unitary vector bundles on a compact Riemann surface*, Ann. of Maths. **82** (1965) 540–567.

[50] Newstead, P. *Introduction to Moduli*, Oxford University Press, Bombay (1978).

[51] Oesterlé, J. *Construction de la variété de modules des fibrés vectoriels stables sur une courbe algébrique*, Birkhäuser **54** (1985) 29–48.

[52] Ramanan, S. *The moduli spaces of vector bundles on an algebraic curve*, Math. Ann. **89** (1973) 14–51.

[53] Schwarzenberger, R.L.E. *Vector bundles on algebraic surfaces*, Proc. London Math. Soc. **11** (1961) 601–622.

[54] Seshadri, C.S. *Fibrés vectoriels sur les courbes algébriques*, Astérisque **96** (1982).

[55] Seshadri, C.S. *Space of unitary vector bundles on a compact Riemann surface*, Ann. of Maths. **85** (1967) 303–336.

[56] Seshadri, C.S. *Mumford's conjecture for* GL(2) *and applications*, Proc. Bombay Colloquium on Algebraic Geometry, Oxford University Press (1969) 347–371.

[57] Shafarevich, I.R. *Basic algebraic geometry*, Springer-Verlag (1977).

[58] Simpson, C.T. *Moduli of representations of the fundamental group of a smooth variety*, Publications Mathématiques de l'IHES **79** and **80** (1994).

[59] Spanier, E. *Algebraic topology*, Series in Higher Mathematics, McGraw-Hill (1966).

[60] Spindler, H. *Der Satz von Grauert-Mülich für beliebige semistabile holomorphe Vektorbündel über dem n-dimensionalen komplex-projektiven Raum*, Math. Ann. **243** (1979) 131–141.

[61] Wu, W-T. *Sur les espaces fibrés*, Publ. Inst. University Strasbourg **11** (1952).

Index

Printed by Printforce, United Kingdom